AGRICULTURAL AUTOMATION

AGRICULTURAL AUTOMATION
PRINCIPLES, SYSTEMS AND APPLICATIONS

Adrian Pegg
Jonathan Chester

Kruger Brentt
Publishers
2023

Kruger Brentt Publishers UK. LTD.
Company Number 9728962

Regd. Office: 68 St Margarets Road, Edgware, Middlesex HA8 9UU

ISBN: 9781787150508

Disclaimer:

Every possible effort has been made to ensure that the information contained in this book is accurate at the time of going to press, and the publisher and author cannot accept responsibility for any errors or omissions, however caused. No responsibility for loss or damage occasioned to any person acting, or refraining from action, as a result of the material in this publication can be accepted by the editor, the publisher or the author. The Publisher is not associated with any product or vendor mentioned in the book. The contents of this work are intended to further general scientific research, understanding and discussion only. Readers should consult with a specialist where appropriate.

Every effort has been made to trace the owners of copyright material used in this book, if any. The author and the publisher will be grateful for any omission brought to their notice for acknowledgement in the future editions of the book.

For information on all our publications visit our website at http://krugerbrentt.com/

PREFACE

Agriculture automation system in the commercial world have developed a new concepts of practical crop management that were not feasible before. Computer-based sensors and actuators such as global positioning systems, machine vision, and laser-based sensors have progressively been incorporated into mobile robots with the aim of configuring autonomous systems capable of shifting operator activities in agricultural tasks. In the application of automatic equipment in agriculture is the use of fleets of robots, in which a number of specialized robots collaborate to accomplish one or several agricultural tasks.

The benefit in this automation technology is enable increase the safety. The development of agriculture was a watershed moment in humanity. Humans' ability to engineer the environment to generate enough food to sustain massive population growth was the first profound change in the relationship between fully-modern humans and the environment. The advent of agriculture kick started a wider range of advancements from the use of fire and prepared food to self-driving machinery. Agriculture has moved us forward us so far in 12,000 years, but we are now at a turning point. And with a global population projection of 9.7 billion people by 2050, agricultural production will need to increase by at least 70% from current levels to serve nutritional trends. Now more than ever, the pressure on farmers to produce nutritious products is putting our planet's health under even more stress. New advancements in technologies ranging from robotics and drones to computer vision software have completely transformed modern agriculture. Farmers now have access to tools that will help them meet the demands of our world's ever-increasing population.

The present book contains fifteen chapters covering all related disciplines. These chapters include Introduction, Trends of engineering systems in agriculture automation, Smart agriculture and vision-based fruit picking robots, Smart farming and precision agriculture, Farm, agricultural land management software, Hardware and components for agriculture monitoring solutions, Software components for managing soil, seeds, irrigation and fertilizers, Aeroponic system, Plant management, Automated weed management and control: a new paradigm in agricultural automation, Automated crop disease monitoring and management, Automation in agricultural equipment and technology, Advances in mechanization and agricultural automation, Global positioning system (GPS) and geographic information systems (GIS) for agricultural automation, Agriculture 4.0: strategic opportunity and challenges Related terminology is given at the end for ready reference. This book provides a comprehensive introduction of automation technologies for agriculture. It elucidates new techniques and their applications in the agriculture sector. Also discussed computerization of agricultural information, types of agricultural mechanism, automation and of the irrigation systems in agricultural sector.

We are grateful to all those persons as well as various books, manuals, periodicals, magazines, journals etc. that helped in the preparation of this book. In spite of the best efforts, it is possible that some errors may have occurred into the compilation and editing of the book. Further queries, constructive suggestions and criticisms for the improvement of the book are always welcome and shall be thankfully acknowledged.

Adrian Pegg

Jonathan Chester

CONTENTS

1

INTRODUCTION

Smart farming (also referred to as digital farming, digital agriculture and precision agriculture) has been proposed to manage land, animals, and farm personnel more effectively. The smart farming approach implies that farm management tasks and upstream interactions in the supply chain are informed by collected data, enhanced by context and situation awareness, and triggered by real-time events. A range of sensors are used to collect these data to monitor animals, soil, water, and plants. The data are used to interpret the past and predict the future to ensure more timely or accurate decision making both on-farm and in the supply chain, where the accumulation of data from different farms also enables so-called Big Data analysis. Scientists and policymakers are increasingly looking to smart farming as a technological solution to address societal concerns around farming, including provenance and food traceability, animal welfare in livestock industries, and the environmental impact of different farming practices

Nowadays, automation plays a major role in various type of industries to minimize the human efforts and increase productivity and achieve customer demand and fulfill customer satisfaction. Today, automation used in the different types of industries like Automobile, electronics, electrical, medical, defense and agriculture industries to improve the quality of products and reduce human errors, cost and time. Most of the industries are affected by the control and uncontrol factors. But for the industrial point of view, there are some limited un-control factors that affect the process because most of the process takes place under manufacturing plants. In the case of the agriculture industry, the output from the agriculture industries mainly based on the different types of aspects that effects on the output directly or indirectly like uncertain prediction of rain, storm, hot and cold waves etc. The conventional agriculture practices is not suitable for this kind of situations. The un-conventional agriculture practices with automation with IoT has been reduce the effects of uncertain factor and help to understand the weather condition that is responsible for increased agriculture output.

Farm automation (or smart farming) is a variety of tech innovations in traditional farming to optimize the food production process and improve quality. As of now, advanced farming technology can be an essential part of the farmer's daily work.

THE IMPORTANCE OF AUTOMATION IN AGRICULTURE

In response to the pandemic, farmers have been caught off guard by this new normal when they are required to adjust quickly. As a result of the Coronavirus pandemic, robotics and automation have become increasingly important in agriculture. In order to ensure food security, agriculture needs to be sustained. In order to enable this expansion and provide food to millions of people, it is most convenient to integrate automation and robotics into farming processes. Agro technology technologies such as harvest automation, autonomous tractors, seeding and weeding, drones, and many others have influenced modern farming worldwide and are meeting the needs of millions of consumers. As a result of automation, there have been fewer labor shortages, more customer preferences, and an increase in crop production.

THE DIFFERENT TECHNOLOGIES IN FARM AUTOMATION

Over the last few years, various novel pieces of technology have emerged in order to assist farmers with their activities. We discuss the most noteworthy technologies related to agriculture automation in this section.

The development and adaptation of various drone and robotic technology have helped farmers tackle the rigorous nature of their profession with greater ease. In particular, the following innovations have emerged in this segment:

Irrigation Drones

The irrigation of fruits and vegetables in farmlands is a crucial step to ensure the health and wellbeing of the produce across their various cycles. Although irrigation is a technique that's been in practice for several decades now, the technologies related to irrigation have stagnated.

This is why drones are being sought after to help reduce the burdens of irrigation for farmers. Drones such as the Agras MG-1 are capable of carrying 10 liters of liquid in a single fill and they can cover over 6,000 square meters of farmland in just under 10 minutes.

Such drones are capable of understanding and adapting to the various specifics of a farm such as altitude, types of plants being grown, and even the weather conditions. When combined with monitoring technology, these drones can identify specific sections of the farm that require greater attention and match these needs with immense ease.

Robots for Planting

Planting crops across large sections of farmlands usually takes several hours of hard labor. This task can also be executed with the help of robots and drones that are capable of planting seeds on the ground and even from the air.

DroneSeed, for example, is a planting drone that's used to help reforest large spaces that have been affected by forest fires. This drone is capable of locating and targeting the precise areas on the ground that are ideal for seeding. They then use compressed air to shoot the seeds onto the ground without allowing them to scatter because of the wind.

ANALYTICAL AND MONITORING TOOLS

After engaging in the planting process and setting up the irrigation channels, farmers spend an enormous amount of time monitoring and analyzing the condition of their crops manually. This is not a precise science and farmers have had to rely on existing information to make hopeful decisions in the past.

Automation in farming can turn this imprecise science into an accurate one with the help of drones that can be continually sent out to monitor farmlands. These drones can be programmed to seek and identify various specifics to help better inform the farmers.

This includes using different sensors and cameras such as infrared technology to be able to identify damaged crops, keep a check on nutritional levels, and even offer a 3D field map to the farmers.

Harvest Automation Tools

The harvesting season is often the most rigorous portion of a farmers' time. A lot depends on this process which is why efficiency and accuracy are the key here.

Harvest robots are therefore being developed to aid in this crucial juncture. Abundant is an excellent example of a harvesting robot that specializes in apples. The use of vacuum technology in this bot has been known to make an enormous difference in its execution.

Agrobot is another similar invention that's used to help farmers harvest strawberries. It uses A. I technology to understand the ripeness of the fruit before it can harvest them.

AUTOMATIC IRRIGATION

Automated irrigation processes can be useful for two specific functions — assisting with subsurface drip irrigation (SDI) systems and offering information with the help of sensors.

SDI technology is useful for providing precise levels of water to necessary plants and controlling the drip. When coupled with IoT sensors, it can help farmers study the moisture levels in the soil at all points and take necessary action when necessary.

Robotics for Seeding and Weeding

Farming robots have been specially created to help with the seeding and weeding process. These robots utilize technologies such as artificial intelligence and computer vision to maximize their efficiency and thereby make full use of the resources at hand.

Even the process of applying fertilizers can be done with these robots, and their accuracy has an incredible impact on the soil and environment of the region.

Robots for planting are focused on the field's specific area, and they work with great precision. This type of farming robot uses artificial intelligence and computer vision, which allows for a reduction of pesticides in the field and, subsequently, the production of high-quality food.

Fig. 1: Seeding and weeding robotics run on solar power

Eco Robotics produces completely autonomous robots that use solar power for their work. This is the greenest approach: a small four-wheel machine is moving across the field and spraying herbicides with minimum damage to the crops and environment.

Harvest Automation

Harvesting is not easy work for machines – they should be gentle enough not to damage fruits and vegetables. Nevertheless, harvest robots already exist and they cope with their tasks successfully.

For example, Abundant is the first commercial company specializing in apple harvesting. The trick is to use a vacuum instead of graspers or similar devices. Agrobot, in its turn, is closer to the ground – the company builds robots to harvest strawberries. This innovation determines the fruit's ripeness via AI technology.

Automated Tractors

Automated tractors have been making waves in the field of agri-tech. These machines can be remotely controlled and can help farmers open up a lot of spare time in order to focus on the more immediate needs of their farms.

Smart tractor technology is certainly on the up and up, and it's expected that these vehicles will become fully autonomous in the near future. This means that they will be able to utilize various ancillary technologies such as GPS and light detection sensors to carry out various functions of the farm without the supervision of the farmers.

A good example of automated tractors comes from Bear Flag Robotics who specializes in creating driverless tractors. Farmers can now remotely plan the routes of these vehicles and even gain real-time reports of their land as the tractor goes about its duties.

Fig. 2: Automated Tractors

DRIVERLESS TRACTORS

The tractor is the heart of a farm, used for many different tasks depending on the type of farm and the configuration of its ancillary equipment. As autonomous driving technologies advance, tractors are expected to become some of the earliest machines to be converted.

In the early stages, human effort will still be required to set up field and boundary maps, program the best field paths using path planning software, and decide other operating conditions. Humans will also still be required for regular repair and maintenance.

Nevertheless, autonomous tractors will become more capable and self-sufficient over time, especially with the inclusion of additional cameras and machine vision systems, GPS for navigation, IoT connectivity to enable remote monitoring and operation and radar and LiDAR for object detection and avoidance. All of these technological advancements will significantly diminish the need for humans to actively control these machines.

According to CNH Industrial, a company that specializes in farm equipment and previewed a concept autonomous tractor in 2016, "In the future, these concept tractors will be able to use 'big data' such as real-time weather satellite information to automatically make the best use of ideal conditions, independent of human input, and regardless of the time of day."

Fig. 3: Driverless Tractors

WEEDING AND CROP MAINTENANCE

Weed and Crop Detection

GPS Guidance. Because of the small size of the crop at the optimum time in the plant life cycle for weed control precise lateral guidance of the sensors and actuators is a critical prerequisite for robust performance in automated weed control systems, particularly those based upon spray targeting of individual weeds. The most commonly used technology for precise lateral guidance in specialty crops is currently the real-time kinematic (RTK) global positioning systems (GPS) technology for autoguidance of the tractor. Most modern RTKGPS guidance systems are compatible with global navigation satellite systems from multiple countries to increase the total number of satellites available and improve reliability.

Weed Control Actuators

These are devices like cultivators that kill weeds by uprooting, thermal weeders like flame, lasers or steam that kill the weeds by destroying plant membranes, abrasives that physically degrade weed foliage, and mowers that cut weeds. Weed detection systems and processors signal the actuator to control the weed but not the crop.

Physical Weed Removal

There are four primary methods of physical weed control that may be automated: (1) mechanical cultivation; (2) thermal weed control; (3) abrasion, i.e., bombardment with air-propelled abrasive grits; and (4) mowing. All of these methods have the advantage of being pest control devices "which work by physical means to control a pest" as defined by the U.S. Environmental Protection Agency (USEPA), and do not require registration. Weed control devices also have the advantage of being organiccompliant and therefore useful in organic settings.

With the exception of air-propelled abrasive grits, all of these physical methods for weed removal have been in use for decades. Automation is a means to take a proven weed control device like a cultivator knife, and combine it with intelligent technology to create something very different—an intelligent weeder.

Intrarow Cultivation

Traditionally intrarow weeds have only been controlled by hand weeding and selective herbicides. Cultivator tools such as finger weeders and torsion weeders are also used, but generally only remove small weeds. Growers and researchers have been searching for a way to remove weeds from the intra-row space mechanically without damaging the crop.

Two designs for intelligent intrarow cultivators are currently on the market: the rotating disc design from Tillett and Hague Technology Ltd. in the United Kingdom and is being marketed as the Robocrop "In Row" cultivator and reciprocating knives that reach in and out of the crop row with the use of machine vision guidance. Both systems essentially do the same thing, but by different mechanisms. The Robocrop cultivator has a rotating disc controlled by a vision system to detect the crop plant and align the disc cutaway section with the crop plant (Fig. 4). The disc rotational phase is altered as needed by changing the speed of the hydraulic drive to align the cutaway section with the crop plant to allow for variation in crop spacing.

Fig. 4: Intrarow cultivator (top) Close up of the intrarow cultivation (bottom)

Research showed that the Robocrop rotating cultivator provided effective weed control and did not injure or reduce yields in transplanted vegetable crops compared to the standard cultivator treatment. However, in seeded lettuce where plant spacing is more variable, the Robocrop injured the crop and caused 26% yield reductions, primarily due to stand reduction.

Examples of reciprocating knife type intelligent cultivators currently available for commercial use include the Robovator and Steketee IC. The Robovator intelligent cultivator was also evaluated in broccoli and lettuce in comparison to a conventional interrow cultivator. The Robovator reduced the weed densities by 27 to 41% more than the standard cultivator and hand-weeding times by 29

Fig. 5: Intrarow cultivator equipped with reciprocating knives, weeding lettuce

Fig. 6: IC, equipped with reciprocating knives operating in a fennel

to 45%. They also found that broccoli and lettuce yields were not reduced by the Robovator. Another commercially available intelligent cultivator is the Steketee IC cultivator. Evaluations of the IC cultivator in lettuce showed that the machine found the crop row readily and detected the crop with near 100% accuracy early in the crop cycle, but fell to 54% accuracy at crop maturity.

Intelligent cultivators were found to be a viable alternative to hand weeding for vegetables. Robovator intelligent cultivator was evaluated in onion [Allium cepa var. cepa (L.)] and cabbage [Brassica oleracea (L.) var. capitata (L.)] for intrarow weed removal in comparison to a finger weeder. They did not see a large advantage for the intelligent cultivator compared to the "nonintelligent" finger weeder in terms of weed control and crop tolerance.

Another concept for controlling intrarow weeds mechanically is "stamping." Stamping is where weeds are pushed into the ground with the use of a high-speed "ramming rod." Researchers developed one such system for use in organic carrot [Daucus carota (L.) var. sativa (Hoffm.)] production. The stamping tool comprised of a cordless nail gun mechanism, which forced a 1.0-cm-diameter cylinder rod 30 cm into the soil. The stamping tool was attached to a robotic arm that allowed the device to move in the X–Y plane for precision placement. The robot was equipped with two cameras, one mounted in front of the vehicle for detecting the general location of weed plants and the second for precise positioning of the stamping tool. Although actuator positioning was generally better than 5 mm and stamping tool cycling speed was 100 ms, the maximum work rate of the system was only 1.75 weed/s–1 at a speed of 3.7 cm/s–1.

Thermal Weed Control

Flame weed control, used in organic and conventional cropping systems, is a process of exposing plant tissues to flames coming from a burner normally fueled by propane. Flaming controls weeds by heating rather than burning plant tissue. Propane burners can generate combustion temperatures of up to 1900 C, which raises the temperature of the exposed plant tissues rapidly. Heat injury results in destruction of plant membranes, which results in loss of cell function, and eventually the plants die or are severely weakened. During the 1960s, flaming was widely used in the United States for weed control in cotton (Gossypium hirsutum L.), maize (Zea mays [Schrad.] Iltis), sorghum [Sorghum bicolor (L.) Moench], soybean [Glycine max (L.) Merr.], potato (Solanum tuberosum L.), and other crops. During the 1960s and the 1970s, flaming was replaced by the use of herbicides because of an increase in propane price and the availability of less expensive herbicides. Recently, concerns about herbicide impacts on the environment have renewed interest in flame weed control.

The use of automated flamers for intrarow weed removal remains an area to be explored (Nørremark et al. 2009). A prototype automated flamer for intrarow weed removal is under development. The basis of this machine is serial application of a flame fueled by propane. As the flamer passes over a weed, the machine vision system turns on the burners so that a specific weed will receive multiple exposures to the heat. The machine vision system turns off the flame as it passes over the crop. The potential for use of precision steam application to control weeds either by precision foliar application to control weeds or to disinfect soil killing weed seed prior to crop planting remains to be explored.

Hot Oils

A robotic system for precision, pulsed-jet, microdosing of high-temperature, organic, foodgrade oil for intrarow weed control in vegetable crops was developed at the University of California, Davis. This robotic system has two unique features. First, it was capable of distinguishing tomato plants from weeds, including black nightshade, with above 95%

accuracy with the use of an advanced sensing system that used hyperspectral imaging and a Bayesian classifier based upon machine learning for real-time species recognition. Second, the robotic pulsed-jet, microdosing system (the actuator) could target individual weed leaves with a lethal dose of high-temperature (160 C) oil with a 1-cm spray resolution. The target application rate to weed leaf surface was 0.85 mg cm2 of the thermal fluid, applied in 10-ms-pulsed doses, which achieved above 93% weed control at 15 d after application with less than 3% of tomato plants injured by the treatment. Commercial development potential is unknown.

Laser Weeding

Weed control using laser technology has been investigated by several researchers as a nonchemical/organic weed method for management. Researchers studied the potential of commercially available laser systems for controlling three weed species: common chickweed (Stellaria media L.), scentless mayweed [Tripleurospermum inodorum (L.) Sch. Bip.], and oilseed rape (Brassica napus L.). Two different types of continuous-wave diode lasers were used to target the apical meristems of weeds at the cotyledon stage. Weed control efficacy varied significantly and depended on weed species, wavelength, exposure time, spot size, and laser power. Of the systems tested, only the 5-W, 532-nm laser with 1.8-mm spot size configuration effectively controlled all three weed species. Energy and exposure times required for this configuration ranged from 1.3 to 9.9 J, with corresponding exposure times of 250 to 2,000 ms. Further research is needed to determine the efficacy on a broader spectrum of weed species.

Researchers also experimented with using lasers for nonchemical weed control. They investigated four types of lasers at different wavelengths: CO_2 (10,600 nm), fiber (1908 nm), diode (940 nm), and solid-state (532 nm). The lasers were used to irradiate the apical meristem of redroot pigweed (Amaranthus retroflexus L.) at the first true leaf stage of growth. Three different laser positions and three spot sizes 3.0, 4.2, and 6.0 mm were used to investigate the impact of local or systemic irradiation on plant damage. Minimum lethal doses of energy ranged from 10 to 71 J weed1 with the lowest energy requirements resulting from use of the CO_2 laser. Laser positioning was found to be important, as a damage model showed a required energy increase of 1.3 J for every 1% loss in positioning accuracy.

Abrasion

Air-propelled abrasive grit material has been evaluated as a weed control device directed at the base of corn, soybean, or vegetables. Similar methods have been evaluated in the Netherlands with pneumatic air blasting. Currently there is research to pair abrasion with intelligent technology, but to the best of our knowledge there are no published reports yet.

Mowing

The main area where autonomous mowers have been utilized is for mowing lawns. Traditionally, autonomous lawn mowers have utilized a wire guidance system that sets the boundaries for the area to be mowed. However, a project to develop highly accurate GPS guidance for autonomous lawn mowers is in progress in Europe. Traditionally mowing has been used for weed control in orchards, vineyards, and pastures. There would appear to be potential for use of autonomous mowers for weed control in all of these areas. There has been some research in Denmark to develop an autonomous mower for weed control in Christmas tree plantations. It would appear that the potential exists for further development of autonomous mowing vehicles for several applications in orchards, vineyards, and pastures.

Weeding and pest control are both critical aspects of plant maintenance and tasks that are perfect for autonomous robots. A few prototypes are already being developed, including Bonirob from Deepfield Robotics, and an automated cultivator.

The Bonirob robot is about the size of a car and can navigate autonomously through a field of crops using video, LiDAR and satellite GPS. Its developers are using machine learning to teach the Bonirob to identify weeds before removing them. With advanced machine learning, or even artificial intelligence (AI) being integrated in the future, machines such as this could entirely replace the need for humans to manually weed or monitor crops.

THE ADVANTAGES OF FARM AUTOMATION TECHNOLOGY

The green revolution forever changed the nature of farming for decades to come, and the smart revolution can have a similar impact on the industry as well. We look at some of the major advantages of automation in farming in the section below.

Meeting Growing Needs

The agricultural needs of the planet are truly immense at this point, and current farming techniques will not be able to meet these demands. With smart agricultural farming techniques, however, it's possible to maximize the potential of farmlands, minimize the stress on farmers, and heighten the efficiency of current processes. Over time, these methods will be able to match the needs of our population.

Solving Labor Shortage Problems

As a bulk of all farming processes are currently labor-intensive, it requires a steady supply of workforce around the year. Unfortunately, farming is not a lucrative field which is why there is a severe shortage of laborers in this segment.

Robots and drones can be used to match this dearth of laborers by taking over menial and repetitive tasks and freeing up the workforce to focus on more lucrative professions.

Caring for the Environment

The improper use of chemicals on farmlands and the continual encroachment of forests to convert them into farms is taking a toll on the environment. Efficiency is the key here to solve these issues. It's perfectly possible to conduct farming on existing spaces without cutting down more forests. All it takes is for smart farming techniques to maximize the utility of space and improve existing processes to protect the environment.

Reducing Labor, Increasing Yield and Efficiency

The core concept of incorporating autonomous robotics into agriculture remains the goal of reducing reliance on manual labour, while increasing efficiency, product yield and quality.

Unlike their forebears, whose time was mostly taken up by heavy labor, the farmers of the future will spend their time performing tasks such as repairing machinery, debugging robot coding, analyzing data and planning farm operations.

As noted with all of these agbots, having a robust backbone of sensors and IoT built into the farm's infrastructure is essential. The key to a truly "smart" farm relies on the ability of all the machines and sensors being able to communicate with each other and with the farmer, even as they operate autonomously.

DRONES

Drones have an important role in agriculture's innovative changes. Earlier, having a glance at a field was possible only with helicopters or even satellites – things have changed now. Using a drone is cheaper and does not require any special human skills like flying a helicopter.

Drones can take aerial photos and videos, and the files' quality is getting better with time. But this is far from the only way to use drones in agriculture.

Monitoring and analytics

The average field size makes it almost impossible to properly monitor the state of a harvest. Previously, the most reliable solution was satellite imagery, but still, it did not give enough guarantees to farmers: such images were inaccurate and did not provide real-time information.

With drones, automation in farming is brought to the next level. Drone devices can identify damaged crops using different types of light, including near-infrared. Farmers receive this information immediately and can quickly react to the changing conditions.

Drones are also useful when creating a planting scheme. They provide 3D field maps with soil analysis – it helps farmers make the right decisions at the very beginning of the crop cycle.

Real-Time Monitoring and Analysis

One of the most useful tasks drones can take on is remote monitoring and analysis of fields and crops. Imagine the benefits of using a small fleet of drones instead of a team of workers spending hours on their feet or in a vehicle travelling back and forth across the field to visually check crop conditions.

This is where the connected farm is essential, as all this data needs to be seen to be useful. Farmers can review the data, and only make personal trips out into the fields when there is a specific issue that needs their attention, rather than wasting time and effort by tending to healthy plants.

Given that drones for agricultural use are still early in their evolution, there are a few downsides. Ranges and flight times are not as robust as many farms would need—currently, even the longest running drones max out at around an hour of flight time before needing to return and recharge.

The capital expenses are also still quite high, as much as $25,000 USD per drone for something like the PrecisionHawk Lancaster. Less expensive models exist, but they may not come with the necessary imaging or spraying equipment.

PLANTING

Automated farms can use robots for seeding, not only on the ground, but also for planting from the air. Let's see how it works.

The main goal of DroneSeed is to help reforest lands after wildfires. The drone carries a module that "shoots" seeds at the set place and specific distance from each other. To make it possible, developers use compressed air – this way, seeds will not scatter around the field

However, is not the only branch which requires the help of a drone. We might see "air planting" used more often as a part of agriculture automation.

Principles

Ecosystem Principles

- **Ecosystem Thinking:** Design all digital initiatives as ecosystems, and NOT as systems.
  - Ecosystems span across Centre and States, public and private, and are composed of several autonomous, interoperable, and federated systems.
- **Building Block approach:** Architect and design systems and ecosystems in terms of minimal and reusable Building Blocks.
  - Categorize the Building Blocks as Core, Common or Reference Building Blocks basing on the degree to which they are to be decentralized. All the building blocks shall be able to evolve orthogonally, thereby meaning that any building block can be modified without requiring the other building blocks to make any consequential changes.
- **Open API-based:** Adopt the principle of 'Open-API by default'. Exceptions shall be justified.

- **Open, Open and Open:** Design the digital systems to be built on open source, to be published as open source, and to conform to open standards. Exceptions shall be justified.
- **National Portability:** Design national digital systems and platforms for portability across India.
  - Factor the requirements of localization and diversity, inclusion, and special needs.
- **Participatory Design:** Drive participatory design and end-user engagement at all stages of the digital initiative.
  - To enable this, establish a set of robust and transparent rules of engagement for the ecosystem players.
- **Innovation:** Enable and promote innovation, and 'responsible' deployment of emerging technologies. Establishing an 'ecosystem sandbox' can be good enabler of open innovation.

Architecture Principles

- **Federated Architecture:** Adopt a Federated Architecture model for designing digital ecosystems – especially data and applications.
  - Architect around the constructs of Single-Source-of-Truth, System-of-Records, and 'unbundled applications'.
- **Agile Architecture:** Permit the architecture to evolve in an agile and iterative manner.
  - Do not restrict or constrain the potential by being prescriptive when not necessary. Adopt the principles and methods suggested in Agile IndEA Framework.
- **Technology Independence:** Architect systems to be technology independent.

Business (Domain) Principles

- **Value-driven:** Focus on providing additional or new value to the user. Plan to Define, Design and Deliver value.
  - Measure service levels (planned and achieved), impact and outcomes.
  - Adopt the methods suggested in Digital Service Standard.
- **Integrated Services:** Identify, design, and deliver integrated services that cut-across agency boundaries - public and private, to realize the goal of Connected Government.
- **Outcome-driven:** Define service levels and outcomes benchmarking with the best and build services around such outcomes. Work backwards to reengineer the processes - where necessary.

- **Choice:** Provide choice, by design, to the farmer.
- **Universal Access:** Ensure that the digital services are accessible to all the target groups, including those residing in remote areas and to the disadvantaged groups. Ensure that the benefits of IDEA reach small and marginal farmers.

Technology Principles

- **Cloud First:** Adopt the principle of Cloud First or Cloud-by-default and leverage the full range of benefits offered by the cloud technologies.
- Mobile First: Design the delivery of all digital services through mobile by default.
- **Data ARE assets:** Design data systems in a manner that creates, supports, maintains, and enhances value to the ecosystem. Promote establishment of Data Exchange(s) that enable regulated exchange of data for public purposes, innovation, and research, and for permissible commercial purposes.
- **Data sharing:** Lay down clear policies specific to the agriculture domain that enable and regulate the sharing of data, in conformance with the applicable data protection regulations.
- **Standards:** Specify the existing technology and data standards applicable to the ecosystem and define methods to ensure compliance with the same.
- **Privacy-by-Design:** All entities participating in IDEA shall ensure to protect the personal information and publish a privacy policy(ies) that conform(s) to the principles of Privacy-by-Design.
- **Security-by-Design:** All entities participating in IDEA may design and enforce a cybersecurity policy that conforms to the principles of Security- by-Design, and an ISMS (Information Security Management System) that conforms to the ISOs relating to information security.

THE CHALLENGES OF FARM AUTOMATION TECHNOLOGY

As robotics and drones are relatively new inventions being adopted for agriculture, it's perfectly understandable that the current cost of acquiring these tools is high. This is also one of the biggest challenges of farm automation as a vast majority of farmers cannot currently afford such techniques.

It comes down to governments and corporations working together in order to make these technologies more affordable for farmers. The research and development that needs to go into this segment are on the higher end for now, but it can have profound implications on the future of our planet.

Several obstacles stand in the way of farm automation. As a result of the high costs of adopting robotics, farmers have the biggest barrier to entry, especially in poor countries. Generally, robotic planters must be designed in a way that can carry large quantities of water and pesticides, which results in additional costs. High repair costs are also a consequence of equipment breakdowns and various technical concerns. For farmers to be able to fully exploit farm automation, new technologies must be combined with their knowledge and experience.

Agricultural automation and robotics: challenge

In planting process, numerous works have been developed to design an autonomous planting process with great accuracy. Based on the conducted review, most of the researches are focusing on the seed uniformity and detection which is one of the main objectives of planting process. However, more focuses should be also concentrated on the path correction to maintain the straight-line row.

Challenges such as non-uniform soil surface and soil stiffness for a diverse type of soil such as sand, loam and clay with dry or muddy soil condition has made the research to become more challenging to track and maintain the robot structure in executing planting task thus sometimes it may lead to inconsistent row arrangement.

For inspection process, most of the designed algorithms were able to identify the disease or defect in simulation and real experiment with a great degree of accuracy. However, most of the implementations in real experiment are executed by taking a real plant disease or quality defect and it was then placed into a clean background for detection purpose. This procedure must be further improved by taking the real plant image inside a real environment with dynamic background. Therefore, the plant disease and quality defect detection will become more realistic to be implemented for future robot design thus the robot would be able to clearly imitate the way human conduct the plant inspection process.

Based on the conducted review for implementation of robotics and automation for spraying task, most of the implemented researches are focusing towards designing an efficient spraying system with minimum operational costs. In addition to the conducted researches for spraying, extensive focus also needed for spraying management where the developed autonomous system must be able to identify an optimized route to execute the selective spraying operation by considering several spraying characteristics such as spraying capacity and refill mechanism to ensure that a fully autonomous system to be realistically implemented in real spraying application. In harvesting, most of the conducted researches are focusing on the target detection to identify the location of the harvested product. It is undeniable that the target detection is one of the most important research areas in harvesting operation however, more researches can also be conducted in harvesting management where the autonomous system is able to plan a specific strategy to harvest a maximum amount of agricultural product with shortest time and at the same time considering the constraints of storage capacity and the distance to the depot. Therefore, the harvesting process can be executed efficiently with a minimum operational cost.

Despite several challenges in different agricultural operation, farmers are also concerned about the cost requires to invest into agricultural robotics and automation. Some of the farmers are afraid to invest their money into technologies that may not benefit them in the future. Therefore, agricultural researchers need to come out with ideas in designing a multifunctional agricultural robot with affordable cost. One of the suggestions that may be used in designing agricultural robot is the modular robotic design with great robustness. As different operations in agriculture have different working mechanism and method, it is recommended that the structure of the robot to be easily detachable and can be replaced by other structures based on the specific function and requirement of operation. Therefore, farmers only need to invest to one main robotic system with different types of detachable structure to complete the agricultural operations from planting to harvesting. From that, the investment cost will be much lower as the farmers are not required to buy different types of robot for different agricultural operation.

Smart Decisions, Brighter Futures

Smart farming is here to say. In fact, we can be bold enough to say that it's the only way forward. With population estimates expected to hit the 10 billion mark in the coming days, there is simply no alternative but to adopt technologies to maximize farming techniques and make way for fully autonomous agricultural production.

BENEFITS OF AGRICULTURE AUTOMATION

One agrobot can change the future. Smart farm automation technology has great potential to be tremendously impactful. The solution to numerous global problems and the creation of eco-friendly approaches lies in this field of technology.

The agriculture industry meets consumers' demands.

We have mentioned that consumer demands and their preferences are changing. We all want food to be fresh and be quickly delivered to shops and markets. Agrobots allow farmers to save time and money.

The labor shortage in the industry is decreasing.

Labor in some farming branches can take up to 50% of the overall cost. The labor shortage makes this situation even more dramatic. Using robots for seeding, harvesting, watering, and monitoring decreases the need to hire employees for numerous monotonous tasks.

Agriculture is becoming more eco-friendly.

The usage of pesticides frightens lots of people, that is why agriculture is gaining a infamous reputation. Robots are shifting the rules of the game towards an environmentally-friendly approach: they can add pesticides with great precision, which is almost impossible when done by an actual human.

The revolution in agriculture is around the corner

Several years ago, seeding drones were only a phantasy of several brave developers – and look where we are now. Robots and drones work together, creating connected smart farms. They exchange information, send signals to each other, and help agriculture to be ten steps ahead and feed milliards of people.

But any IoT solution has software instruments that stand behind and support the farm, no matter how complex it is. Every drone or automated tractor needs a brain to make the right decision about the plants in the field.

Challenges of farm automation

Still, there are challenges of farm automation that must be overcome. High costs to adopt robotic technologies presents a huge barrier to entry for farmers, especially in developing countries. For example, robotic planters must carry water or pesticides with significant weight; the hardware must be built differently, which results in higher costs to make it a larger size. Technical issues and breakdown of equipment also present high costs to fix for such specialized equipment. In order to fully utilize farm automation, farmers will need to combine their knowledge and experience with these new technologies.

2

TRENDS OF ENGINEERING SYSTEMS IN AGRICULTURE AUTOMATION

Modern-day farming and agriculture are no more a reminder of the laborious sweat-intensive field-work for farmers and slow growth patient process. With the advancement of agriculture trends 2022, the past decade has seen a thriving hike in agricultural yields, excellent examples of contemporary farming, and many more.

TECHNOLOGY

Although the current technological developments have reached an astonishing point, it is difficult to obtain continuous and daily results due to the narrow margins, high costs and limitations of use. With new developments in artificial intelligence, machine learning, sensors, IoT (Internet of Things) and data management, these technologies will be commonly used daily at lower costs. According to the results of the Global Artificial Intelligence Analysis in the Agricultural Market Analysis conducted by BIS Research, it is predicted that the market will grow by 28% in the 2019-2024 period.

AUTOMATION

Thanks to broadband access in rural areas and the increase in technological informations, automation will be used in equipment, implementation and decision-making over the next decade and it seems to be revolution. The agricultural IoT market, which was recorded as $11.2 billion in 2018, is expected to reach $20.9 billion by 2024.

Apart from autonomous cars and tractor models, it will be possible to see a wide range of automations in the coming years: Automatic irrigation systems based on weather and crop conditions, automatic storage fans and temperature controls, early warning/alerting systems for potential hazards to production or marketing, automated job creation etc.

AGRICULTURAL REVOLUTION

Today, farmers worldwide are adapting to the technical aids to improve their farming efficiencies and gain better yields. Lance Donny, the new technology agriculture leader, says that we need to increase crop production with scarce resources-land, water, and fertilizers. Some of the state-of-the-art farming tools and technologies are crop monitoring drones, livestock measuring sensors, farm management software, driverless tractors, and many more.

Agriculture has drastically changed since the birth of human civilization. It has moved from being a labor-intensive industry to a smart and logical decision-making one. Today's farmers are educated, and they make intuition-based decisions. However exciting this might sound, there's a hidden risk behind it. To cost-effective and proven solutions, farmers must use the generous amount of data available on the internet.

We have seen the magnificent galore of trending new technology in agriculture. Be an early adopter of these technologies to bring a drastic improvement in your agriculture business.

FIVE DISRUPTIVE AGRICULTURE TECHNOLOGY TRENDS

Let's have a look at emerging trends in agricultural technology for 2022.

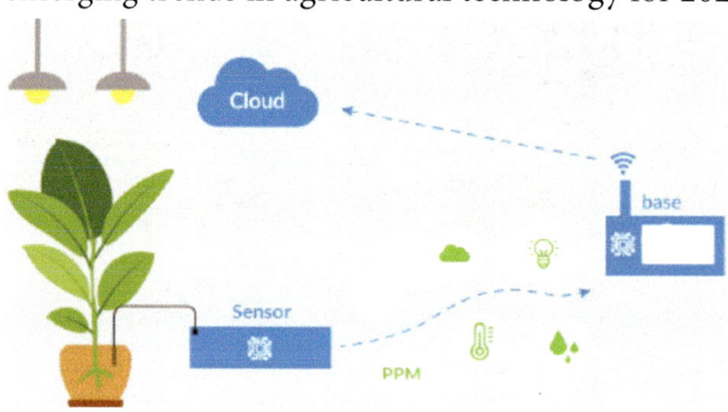

IoT in Agriculture

The use of the Internet of Things in intelligent farming is aided with various sensors implanted in the agricultural farm. The different sensors used are light, humidity, soil moisture, temperature, crop health monitoring, etc.

Some of the prime use-cases of IoT in Agriculture are:

- Data Collection by farm sensors like autonomous vehicles, wearables, button cameras, robotics, control systems, etc.

- Aerial and ground-based drones for irrigation, assessing crop health, spraying, monitoring & field analysis.
- Geofencing using wireless IoT sensors and livestock tracking to monitor cattle healthcare.
- Predictive analytics for rainfall, temperature, soil, humidity, etc.
- Innovative Greenhouse with the aid of IoT devices and monitors, which doesn't require human intervention.

Farmers have accounted that IoT is a cost-efficient way to increase agro production, and hence agro-entrepreneurs like yourself must invest in fascinating IoT app development for your business.

Geographic Information System (GIS) in Agriculture

GIS is a technology that represents any geographical entity in the spatial representation using hardware, software, and data. The hardware used in satellites, drones, GPS systems to locate data points and fetch information from them for analysis.

In the agricultural domain, farmers can use GIS to analyze complex spatial data like rainfall amount, topography, soil elevation, slope aspect, wind direction, flooding, erosion, etc., and so much more. Several satellites are already launched by the government or universal bodies, e.g., Landsat 8, where you need to pay a specific fee for accessing your geography data.

Some excellent use-cases of GIS in the agriculture industry trends 2022 are:

- Irrigated landscape mapping
- Crop health assessment
- Irrigation amendment analysis
- Land degradation assessment studies
- Erosion remediation
- Efficient drainage elevation models

GIS and its generous usage have led to its other names- satellite farming or precision agriculture for all the wonders that offer. Moreover, with the advancements in GPS, robotics, and unnamed aerial vehicles, various farming operations are now computerized.

AI/ML & Data Science in Agriculture Technology

Artificial intelligence is the application of human intelligence via a machine body where instructions are fed into the machine, and hence you get clever work without effort. The entire lifecycle of agriculture involves preparing the soil, seed sowing, adding fertilizers, irrigation of water, protections from weed, harvesting, and storing. At every stage, growers or farmers need to rely on their instincts, calculations, and risks based on the right time and other factors. AI and ML can contribute devastatingly and benefit from their proven data analysis and predictions.

All the crucial agricultural data collected by IoT devices and ML algorithms are processed and channelized with data science. Farmers cannot use the raw data, and hence data science is changing farmers' lives in making vital decisions.

Real-time use cases of AI/ML and data science in agriculture are:

- Predicting yield and quality assessment
- Predictive analytics for crop sustainability
- Using ML to eliminate weed by recognizing species of plants/crops
- Detection of crop infections and diseases
- Intelligent harvesting & pricing decisions
- Prevention of wastage and meeting demands
- Autonomous robots for herding cattle

Blockchain

Once the crops and produce are ready, farmers dive into the troubles of fair trading, selling, marketing, and proving the authenticity of their produce. Hire Blockchain developer helps farmers ensure the safety of their crops, preventing theft and fraud, efficiently managing the supply chain, and balancing the food ecosystem.

The real-time use cases of Agriculture Blockchain technology are:

- Food Traceability
- Transparency in the food supply chain
- Agricultural insurance for farmers
- E-commerce for agro-trades
- Agricultural subsidies

Automation

Robotics has evolved handsomely, and we aren't surprised when machines are doing farming jobs. It has acknowledged farm automation, or so-called smart-farming, by easing the workload on human resources.

Thus, you can meet the international promise of meeting the rising population's demand by producing much-needed agro-produce with less human intervention and faster. Drones, custom tractors, watering motors, harvesters, and more modern-technology blessings for agriculture automation.

Regenerative Agriculture

Regenerative agriculture lays attention on the rejuvenation and biodiversity of soil and soil distribution. It revives the soil for the upcoming cropping period.

Farmers undertake the following practices while implementing regenerative agriculture:

- Reducing tillage
- No-till farming

- Crop rotation
- Planting cover crops to retain soil fertility

These methods enable sequestration and make your fields act as a carbon-sink.

Controlled Environment Agriculture

These days conventional farming practices aren't easy to attain owing to the climatic fluctuations and extremities. Also, with growing urbanization, growing crops in populated cities along with population hike isn't favorable.

A solution to the above situation is controlled environment agriculture (CEA) where we subject the plants to a controlled potion of light, humidity, nutrients, and water. Some of the planned indoor settings available to attain CEA are vertical farming, indoor farming, greenhouses, etc.

A few new techniques are found to attain a balanced equilibrium for plants-aeroponics, hydroponics, aquaponics, and more.

USE CASES OF MODERN AGRICULTURE TECHNOLOGY 2022

Exclusive use-cases of Automation in Agriculture Technology are:

- Agrobot is a real-time example that harvests strawberries, accessible from a mobile platform, by meeting farmers' standards.
- Apple harvesting using vacuum by Abundant Robotics.
- Autonomous tractors that are programmed in advance to perform driverless control.
- Using computer vision for seeding and sprinkling pesticide where required.

LINEAR ACTUATORS IN AGRICULTURAL MACHINERY AND ATTACHMENTS

The following tasks typically occur in an arable farm over the year:

- Tillage
- Sowing
- Stand care (weed and pest control, fertilization)
- Harvest
- Transport / storage

These tasks are implemented in the combination of tractors with suitable attachments or with a specially made self-propelled machine such as a combine harvester. Most of the attachments are only used for a small amount of time each year, which means that they should save the user money. For example, the cost of deployed mineral fertilizers is many times higher than the initial cost of the spreader itself. Large and expensive self-propelled machines such as combine harvester or field chopper are only purchased by larger companies. For medium-sized and small businesses, these are booked and provided through machinery contractors or machine pool. The individual tasks of an arable farm and the typical tasks and optimization parameters are described below:

Tillage (T)

In the case of tillage, the focus is on the uniform and, depending on the target, specific processing of the area. Typical optimization parameters are the processing quality (e.g. crumbling and soil loosening), the minimization of the energy input and the optimization of the speed. In addition, soil erosion due to processing adapted to the site must be minimized. In practice, the arrangement of the processing equipment into its position, angle and depth is set manually. The depth guidance is controlled mechanically via rollers and trailing wheels. In order to ensure that the tools ease off in the event of obstacles (e.g. stones), the tools usually have spring suspension. The spring strength, the tools and the arrangement are manually adjusted to the specific location. Only the lifting and digging is controlled by the rear power hitch of the tractor. However, there are first machines with "Section Control" for the insertion and removal of individual plowshares and the automatic adjustment of cutting widths and row spacing. Likewise, the automatic and site-specific adjustment of the tool pressure is possible.

Sowing (S)

When sowing, the influencing factors such as the sowing depth, the row spacing and the singling or even distribution of the seed are important. If these are guaranteed, the focus will be on a conventional attachment to allow more area to be processed in less time. Standard machines usually only allow a manual parameter setting. Changes in sowing strength are now also automatically adjustable on new machines. Tillage is often combined with the sowing. The no-till and strip-till techniques are interesting variations. The latter works only on the area of sowing and leaves the rest of the field untouched. Here, too, the desire is to be able to flexibly control the influencing factors specified without having to carry out a lot of manual adjustment work.

Stand care (SC)

All work which serves the preservation and the care of the stand after its establishment is summarized under stand care. This includes fertilization, weed control, and pest and disease control. Different requirements arise depending on the task and the stock. Basically, it is important to act in a particularly plant-friendly manner and to carry out the process optimally and precisely. Precision agriculture offers the greatest potential here.

On the one hand, this is due to the possible material and cost savings and on the other hand to the legal regulations which mainly regulate this area of agriculture (plant protection regulations, fertilizer regulations, water protection, insect protection, etc.). The variable application of mineral fertilizer was one of the first practical precision agricultural applications. Other examples are active field sprayer section control and control of hoeing equipment. In addition, this area covers the largest period of time over the year. While sowing and harvesting have only small working windows, stock management must be carried out and adapted over several months (in small distributed time windows). Likewise, correct stock management improves yield and quality by supporting the stock in its development (e.g. through weed control, pest control and fertilization).

Harvest (H)

Basically, it is important to harvest as much area and crop as possible in a short window of time/weather when harvesting in arable farming. In addition, the harvest should be collected and transported in good quality. For these reasons, harvesters are the largest and most expensive machines in agriculture. For some time, these have already reached the maximum permitted sizes for road traffic. The main focus for development in recent years has been on further optimizing crop quality (sensor-assisted machine setting) and minimizing harvest losses. Further focal points are the reduction of damage caused by ground pressure and the use of driver assistance systems. At harvest time, some of these machines are operated around the clock in order to recoup the high machine costs. For this reason, higher demands are placed on durability and robustness (ultimately reliability) than in other areas.

Transport/Storage (TS)

Especially during harvesting, transportation is essential for the smooth running of a harvest chain. Usually several vehicle combinations are required to transport the harvest from the harvesting machine. There is potential for optimization in approach, overloading time, overloading position and loading quality. For example, when chopping maize, the tractor with loader wagon must be able to drive parallel to the forage harvester during the overloading period so that no crop is lost. This can partly already be done with sensor support, e.g. by means of a camera. The "machine to machine" communication for controlling the tractor speed is also already implemented during the overloading period. Ingredient measurements using near-infrared spectroscopy systems can now be used to determine the dry matter and nutrient content on a site-specific basis. This can be used for billing and storage optimization.

TRENDS IN THE AGRICULTURAL MACHINERY OF TOMORROW

Machine sizes and automation level

In the near future, many farmers will probably continue to rely on existing technology and only slowly use new systems such as robotics, remote sensing using drones or cloud-based IT structures. However, social change can very soon lead to a situation in which only robotics can maintain economic viability. Due to the fact that chemical pest and weed controls may soon be required by law, more crossings with chippers or similar will be necessary in the future. If the trend continues and the effectiveness of a farmer must increase further, this cannot be achieved by even larger machines, but only by partial automation and optimization of the tasks.

The current machines have reached the maximum size to be driven and transported on public roads. It is therefore to be expected that in the future the machines will not be any larger. When robots take over the tasks, the absolute size of the units only depends on the technical possibilities such as the range, the running time and the controllability of the machines. It is to be expected that soil cultivation requires the highest power requirement and therefore defines the lower possible limit of the machines. However, miniaturization will not work up to the level of a garden mower, as the necessary power, running time and terrain mobility can only be guaranteed at a certain size.

Robotics as a solution is already offered for sale by several companies. These robots can work autonomously, but must be supervised by the user. In order to work with these machines, the fields must be adapted to the characteristics of the robots. This includes e.g. planning suitable routes and providing the infrastructure (e.g. electricity, mobile internet). Purely battery-powered and electrically operated machines are within reach. At the moment, the range of conventional fuel driven machinery is still significantly higher than that of their electrified competitors. However, there are already applications where complete electrification can be implemented. This mainly concerns small plots which are well integrated into the local infrastructure. Basically, it can be assumed that the switch to electric motors will lead to an increase in the efficiency of the machines. However, it is not yet possible to predict how energy storage can be meaningfully implemented, whether with batteries, hydrogen or other synthetic fuels.

It is also to be expected that automation will not only enable machines and robots to copy conventional technology and replace the driver, but also adapt the tools to the size, speed and range of the robots. This depends on the task, power and required safety of the machines. In addition, the use of robots creates new working windows, as the machines can theoretically work continuously, even at night.

Working at night offers considerable advantages for weed, pest and disease control, which can be better harnessed with robotics. Likewise, the probability of coming across people at night is lower, which could prevent serious accidents. The autonomous performance of simple activities at night could also improve the effectiveness between man and machine. For example, the farmer could send the final instructions to the machine at the end of the working day to check the quality of the work the next morning. In order to work effectively, the fields must be adapted to the characteristics of the robots. This includes e.g. the pre-planning of suitable paths, the calibration of field boundaries and the provision of infrastructure (e.g. power supply and mobile internet).

Autonomous driving in agriculture is already possible and feasible today, but it is not without risks. In the future, there will always be a certain probability of malfunctions caused by incorrect operation, external influences, software errors or faulty sensors and mechanics. Therefore, the implementation will be determined by the case law. To a certain extent, the legal framework can be transferred from the field of autonomous driving to agriculture. The current legal situation is such that the protection of people always has priority. Liability is divided between the driver, the owner and the manufacturer. A clear distinction is made here between the reasons why an accident could occur and the question of guilt. The case that there is no driver is not yet included in case law. Therefore, automated systems are legally covered, but autonomous systems are not. It is therefore likely that in the near future only systems monitored by a user will be available. This means that the user must take the machine into the field and not allow it to travel autonomously across the field. The most conceivable solution is a small-scale implementation of machines that have enough power to work and do not cause too much damage in the event of a malfunction. It is also possible to monitor several machines using suitable telemetry systems.

3

SMART AGRICULTURE AND VISION-BASED FRUIT PICKING ROBOTS

INTRODUCTION

In the recent era, precision agriculture plays an important role to maintain the future food security with less labor and energy and at the same time improving the environmental management to ensure a productive agricultural output. The work of precision agriculture focuses on controlling the way the seeds, fertilizers and agrochemicals are applied to the soil and how the harvesting process is executed. The execution of agricultural operation differs based on the type of agricultural land. For agricultural lands, the categorization is divided into four types which are confined feeding operations, cropland and pasture, orchards and vineyards and other agricultural land.

As of today, the implementation of precision agriculture is not being limited to only a specific land type. The development of precision agriculture has been focusing into several areas such as technology, digitalization, impact on society, skills, environment and productivity. In technology, the application in wide range of technologies such as geo-referencing, Global Navigation Satellite System (GNSS), data storage and analysis, advisory system and autonomous navigation has been widely used in the development of Precision Farming (PF). The digitalization of agriculture has been widely implemented in parallel with the development of Information and Communications Technology (ICT)-revolution. Digitalization technologies such as cloud computing and Internet of Things (IoT) have enabled an efficient and precise farming management in the scope of numerous data collection which can be easily to be shared and interpreted. Similar to the way which computers have changed.

Recently, automation in agriculture has becoming a trend where agricultural operations such as cultivation, inspection, spraying, pruning and harvesting are executed autonomously to overcome the labor shortcomings in agriculture. The term automated agriculture is used to relate any parts of the equipment or machine that is designed to remove the manual intervention in agriculture. The work of agricultural automation is mainly focusing on autonomous vehicle applications such as robot or tractor where it is being used to minimize the tough, deadly, risky and long working conditions experienced by farmers and at the same time offers a precise and efficient operation and control system. In addition, the output quality and quantity need to be maintained to be in good quality and consumable for human. Therefore, the development of an efficient automation system in agriculture has becoming an interest in current research in agricultural field nowadays to ensure the sustainability of the food security in the near future.

Agriculture is quickly becoming an exciting high-tech industry, drawing new professionals, new companies and new investors. The technology is developing rapidly, not only advancing the production capabilities of farmers but also advancing robotics and automation technology as we know it.

At the heart of this phenomenon is the need for significantly increased production yields. The UN estimates the world population will rise from 7.3 billion today to 9.7 billion in 2050. The world will need a lot more food, and farmers will face serious pressure to keep up with demand.

Agricultural robots are increasing production yields for farmers in various ways. From drones to autonomous tractors to robotic arms, the technology is being deployed in creative and innovative applications.

ROBOTICS AND AGRICULTURAL AUTOMATION

In agriculture, the automation of specific operations has enabled the farmers to manage the crop production efficiently with less energy and cost. Factors such as lack of agricultural workers in addition to the aging farmer's population and increasing agricultural wage has made the farmers and researchers to play an interest in the development of automation system in agriculture. The implementation and development of agricultural automation has been executed by autonomous robots and agricultural machineries such as tractor which usually attached with cultivator, planter, cultipacker and chisel plow. Figure 1 shows several agricultural robots and machineries which requires automation to enhance the efficiency of the agricultural operation.

The execution of agricultural operation needs to be executed by different robotics and vehicle structure based on the type of land and operation requirement. Different robot and vehicle structure had its own limitation that need to be solved by using other machineries. The robotic structure has limitation to execute extreme operations in agriculture due to its sensitive characteristic towards water and mud. Therefore, the tractor is being used to execute such task due to its great ability to traverse inside the muddy structure and less protection towards electronic circuit. On the other hand, tractor application only limits to a wide area due to its large structure. Thus, the application of the small area needs to be executed by mobile robot. For drone application, it is only applicable to open area and its application would be insignificant to a closed area such as the greenhouse as the probability of collision will be increased. To explore more on present application of automation and robotics in agriculture, the categorization was made based on different agricultural operation.

PLANTING

Planting is a process where the seeds or young plant will be planted into the soil for plant growth phase. This process requires a higher level of precision as different plants require specific distance between each plant to optimize the growth and at the same time maximize the yield. The classical way of planting method requires a farmer to manually insert each seed into the soil. This approach requires a lot of time and energy as this process needs a lot of consistency and precision where it usually covers a wide agricultural area. Therefore, a planter machine has been introduced where the farmer will operate the machine by controlling the machine motion and at the same time the seed will be planted into the soil. Figure 1 shows the planters which have been designed for planting process for different plants. From Figure 1, the designed planter usually will be towed behind a tractor where the planter will exert a repeated motion to plant the seed. As the tractor and planter is manually operated by human, the consistency of the row will be affected as the row will not be in a straight line and there are some areas where the planter is unable or miss to plant the seed. Therefore, an efficient autonomous system is needed where it will ensure to produce a straight-line plant row and will not miss any seed planting.

Fig. 1: Planters

(a) Single-seed corn planter,(b) Minimum-tillage planter,(c) Billet planter

To solve the problem faced by manual planting method, autonomous system has been introduced for several crops such as corn, sugarcane and vegetable. To design an efficient autonomous system for planting process, several considerations have become the main objectives in the design process. First, the robot or vehicle must be able to move accurately in a straight path despite on the bumpy road in the farm field. This requirement is very important as the consistency of straight seeding process is needed to ensure the efficiency of the automation process in the next planting process such as inspection or harvesting. The next requirement is related to the level of soil moisture which must be considered as it may affect the soil digging process. In the soil digging process, specific seed requires specific digging depth. Therefore, the digging depth must be kept constant by adapting the soil moisture level and soil compaction in calculating the digging force for cultivation process. Finally, the seeding detection is needed where this system will detect the presence of seed when the vehicle is ready to plant the seed. This system is very important to ensure the planter will plant the seed without missing any unintended planting area.

Most of the conducted research in planting process focuses on the development of the autonomous seeding system with the main objective to ensure that the seed will be planted in a consistent distance and depth. An autonomous seeding robot has been designed using Agribot platform. In this development, an Infrared (IR) sensor has been used to check the condition of seed tank and also for row detection and it showed a satisfactory result in terms of the distance accuracy between the seeds. A seed metering unit control system has been designed to enable the planting consistency. In planting process, seed metering device is usually used for metered seeds discharging into the soil. It is usually used to classify the seed into a single or a group of seed before it is being deposited at the predetermined time with accurate intervals. To access the performance of the seed metering device, different speeds and different seed spacing is needed and the planting quality is measured by using plant spacing uniformity, variation among rows, fuel consumption and negative slippage. From the result, it has been shown that the efficient design of seed metering unit is able to provide a better planting quality and approximately 22% improvement in fuel saving.

Other methods to ensure the consistency of the seed spacing in planting operation is by using a global positioning system. The control system for seed and fertilizer has been developed using Global Navigation Satellite System (GNSS)/ Inertial Measurement Unit (IMU) combined sensing technology. In the designed system, the vehicle will execute the operation based on the target application rate of seed and fertilizer by comparing the target parameter to the time and space data obtained from GNSS/IMU. This application has been proven to be efficient with the maximum error of 4.91% at 5.8 km/h. In addition, researcher introduces a Global Positioning System (GPS) based variable-rate seeding control system for maize planter. In this application, the GPS receiver is used to locate the real-time seeding position, travel speed and course angle in a fixed frequency. Then, the data receives from GPS is further processed to obtain latitude, longitude, travel speed and course angle. This implementation is found to be a low-cost system for an efficient variable-seed control system with an average seeding accuracy of 97.64 per cent.

The consistency of the automation system for planting process is really important as it is the first part of agricultural operation. If the operation is executed inconsistently, it will negatively affect the next operations which are solely depending on the plant distribution which has been developed in the planting process. The development of an efficient planter with seed detection will be able to ensure the optimized planting process with low operational cost and at the same time maintaining the consistency of seed distribution throughout the field. Therefore, the automation of planting process would be much efficient and convenient for the farmers in the future with a good planting quality.

INSPECTION

Inspection in agriculture is a process where the plants are being inspected or observed for any diseases or quality defects. In agriculture, plant diseases are primarily responsible for the reduction in production which causes economic losses. As agricultural environment is very dynamic, numerous unexpected and abnormal stress situations such as abnormalities in temperature, humidity, water levels, disease emergence, and pests has affected the plants and its products. If those abnormalities are not managed in a timely manner, severe and irreparable damages may be occurred. To execute the inspection, traditionally farmers will manually observe the abnormalities in plant by using their human vision system. As the age of farmers are increasing in the past few years, the efficiency of inspection operation has been reduced as the quality of the human vision system is depleted with age. In addition, the implementation of automation in agricultural inspection requires a system to replace the ability of human vision to execute the inspection process. Therefore, computer vision has been widely used to replace human vision in performing plant inspection in agriculture.

Computer vision is an advanced technology for image processing that has a propitious outcome and has becoming a great potential to replace human vision in doing some detailed work in inspection process. A computer vision system has been widely adopted in some heterogeneous domains which includes agriculture. In agriculture, it is noteworthy that image processing and computer vision applications have grown due to the reduced equipment costs, increased computational power and increasing interest in non-destructive food assessment methods. Most of the implementations of vision system in agriculture are used for disease detection and some of the implementations is used for product quality checking. Figure 2 shows some of the implementations of vision system in agricultural inspection process.

Based on Fig. 2, several examples of plant inspection are shown where Fig. 2(a)

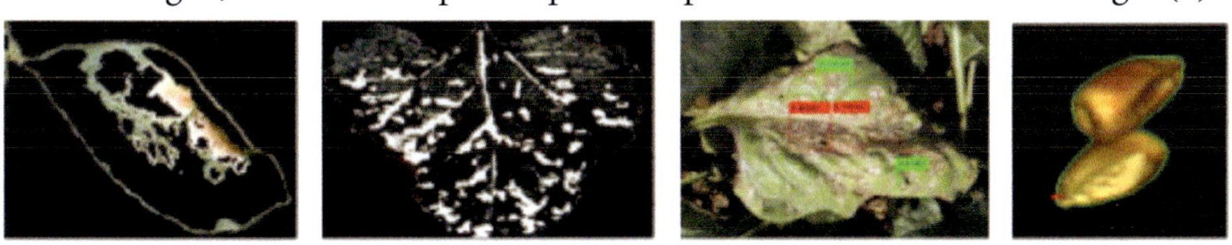

(a) (b) (c) (d)

Fig. 2: Agricultural inspections (a) Rose leaf,] (b) Sunflower leaf, (c) Sugar beet leaf and (d) Corn kernel

shows a disease detection of rose leaf, Fig. 2(b) for sugar beet leaf and Fig. 2(c) for sunflower leaf while Fig. 2(d) shows the quality assessment of the corn kernel. Those implementations are being implemented with specific objectives either to detect diseases or to evaluate the quality of agricultural products. In terms of image processing efficiency, several methods such as Neural Network based algorithms, K-Nearest Neighbors, Support

Vector Machine (SVM), Particle Swarm Optimization (PSO), Genetic Algorithm (GA) and Machine Learning. Most of the implementations of image processing algorithms for agricultural inspection in the past five years are using Neural Network based algorithm such as Deep Convolutional Neural Network, Convolutional Neural Network, Deep Neural Network, Regions with Convolutional Neural Network (R-CNN) and Back Propagation Neural Network [30]. From the results, the Neural Network based algorithm shows a good performance in agricultural inspection with a maximum accuracy of 98 per cent.

In addition of the implementation of image processing method in agricultural inspection, some researchers also make use of the hyperspectral imaging for plant disease detection. A new Normalized Different Spectral Indices (NDSI) is used in detecting peanut leaf spots. This method is being implemented by measuring the deviation of hyperspectral vegetation index which is specific to leaf spot detection. Furthermore, other method such as immunochromatography has been used to detect the fumonisin chemical compound which may contaminate the agricultural product. This method implements an ultrasensitive gray-imaging-based quantitative immunochromatography detection method to trace the chemical compound in agricultural products.

With the advancement of Industrial Revolution 4.0 (IR 4.0), cyber physical system become the main focus area where all devices and sensors are interconnected in a cloud. Therefore, the technology of agricultural inspection is also evolving towards IR 4.0. Recently, IoT are becoming a trend in agricultural inspection process where farmers can monitor and control their agricultural control systems using computer and mobile apps [36-39]. Most implementation of IoT in agricultural inspection are used to monitor plant diseases in real time by capturing images and collecting sensory data such as moisture, temperature, humidity and soil pH level from the farm and displayed inside a website or mobile apps. Therefore, diseases such as early blight, light blight and powdery mildew can be detected in a timely manner by the farmer before the epidemic starts to break. In addition to monitor the farm condition, the IoT technology also enable the farmer to control other agricultural system such as irrigation and spraying and it also able to provide a warning message to the farmers for any abnormalities in plant condition with the help of artificial intelligence.

Inspection process is one of the important agricultural process with the main aim to detect the disease and maintain the quality of agricultural products. The automation of inspection process is really important as it will reduce the inefficiency of manual inspection techniques executed by farmers. The action of autonomous inspection is usually conducted by placing a camera in fixed position, attached on the mobile robot platform or on a drone. From the autonomous design and its implementation in inspection process, the disease prevention and quality checking of agricultural products will become more accurate and efficient to maintain the food security in the future.

SPRAYING

Spraying operation in agriculture is a standard method of applying pest-control chemicals, fertilizers or growing medium which are applied as a fine mist to plants for disease treatment and plant growth management. In most of the farming practices, pestcontrol chemicals are usually applied uniformly throughout the fields to control the spreading of diseases. This technique is applied despite several pests and diseases exhibit at an uneven spatial distribution, especially during early stages of development. Therefore, to minimize the cost of pest-control chemicals usage in agricultural operation, selective spraying was introduced and investigated in the past two decades. The automated selective spraying system, usually executed by a highly automated equipment or mobile robots, enables the selective targeting of pesticide application only where and when it is needed. The main objective of this selective operation is to reduce the amount of pesticide usage and preventing the establishment of infection and its epidemic spread throughout the greenhouse.

In the development of autonomous selective spraying method, previous researches are focusing more towards designing an efficient spraying system with minimum operational costs. To achieve the aforementioned objective, variable rate spraying method is designed that allows farmers to automatically adjust the pesticide or herbicide volume rate to the target based on the canopy size and treatment requirement. Table 1 shows the measured parameters used in designing an efficient variable rate spraying system which is used as a part of autonomous agricultural spraying system in disease control.

Based on Table 1, the parameters chosen in designing a variable rate spraying method differs based on plant species and spraying requirements. Based on those parameters, the volume and the flow rate of spraying agent will be calculated which will be used in spraying operation. The variable rate spraying method is proven to reduce the pesticide use up to 85% compared to homogeneous spray or canopy method. Therefore, this method is highly recommended to be used and implemented by farmers. With the vast automation and robotics technology, this method will be able to reduce the pesticide use with a great precision without exposing hazardous chemical effects directly on farmers.

In addition of reducing the pesticide use, some researches also focusing on navigation management with the objectives to reduce the operational cost of the robot such as time and energy with a great position tracking.

Table 1. Parameters used in designing variable rate spraying method

Disease/Detection	Parameter(s) used
Greenhouse crop diseases	Volume rate, target location and airflow rate
Weed removal in carrot farm	Weed species and size
Vineyard spraying operation	Sprayer travel speed and position (latitude, longitude)
Poplar diseases	Sprayer setting, nozzle types combination, airflow rates and air direction

Disease/Detection	Parameter(s) used
Nitrogen fertilization in greenhouse crops	Plant needs and requirements based on image features (entropy, energy and local homogeneity)
Grape leaf diseases	Average leaf velocity, leaf RMS velocity, turbulence level, spray coverage front/back side and droplet density
Weed removal in wheat farm	Weed spatial distribution
Flower pollination	Gravity, wind and drag
Grape powdery mildew	Level of diseases

This research field is very important to ensure that the robot would be able to navigate towards the target precisely with a minimum travel cost in executing spraying operation. To reduce operational cost, scientists propose a multi-objective algorithm called Nondominated Sorting Genetic Algorithm using Reference Point Based to optimize objectives such as travel time, distance and routing angle. The investigation of robot travel speed effect on mass discharge flowrate of pineapple leaf fibre is conducted where several robot speeds are tested to find the most optimized speed for composite spray operation. Finally, researchers propose a route planning algorithm using Simulated Annealing which considered objectives such as distance, fuel, herbicide volume, input cost and time.

Despite reducing the operational costs in designing navigation system for spraying operation, researchers also focusing on maintaining great position accuracy for agricultural robot. The development of a wheel-type robot tractor is designed for weeding and spraying. To improve its auto steering system with stable navigation accuracy of less than 0.05 m, Real-Time Kinematic GPS (RTK-GPS) and IMU are used for position and attitude sensor for navigation system. As the cost of RTK-GPS is expensive for robot development, a data fusion algorithm using MSPI to filter noisy raw data from low cost sensors such as Differential GPS, IMU and vision sensors for vineyard pesticide spraying robot.

As most of the implementation of autonomous spraying operation in agriculture implement a land-based vehicle, some implementations also consider the use of Unmanned Aerial Vehicle (UAV) for this operation. The UAV swarm is used for non-uniform crop spraying where a multi-agent area coverage method Heat Equation Driven Area Coverage (HEDAC) used to reduce over-spraying. In addition, researchers implemented Genetic Algorithm to adapt its behavior to dynamic weather conditions for pesticide spraying application. From this implementation, the spraying accuracy can be maintained by controlling the behavior of UAV adaptively based on several parameters such as wind speed and direction and pesticide deposition.

The replacement of human force in executing spraying operation is very important to reduce the risk of hazardous chemical substances to the farmers. With the vast development of automation technology, the transition phase of human replacement in spraying operation will become much easier with the main objective to treat the plant disease with minimum operational cost. However, much researches are still being conducted to ensure that the implementation of automation in spraying operation will provide infinite benefits to the farmers in controlling plant diseases to maintain the food security in the future. Therefore, researchers are still able to improve the current methods and techniques in developing an efficient spraying operation in agricultural application.

HARVESTING

In agriculture, harvesting is a work of collecting the agricultural products to be processed or sold. To operate this process, the fruits or vegetables need to be collected and stored for further processing or it may be directly sold to the buyers. As this process need a detailed observation with repetitive procedure, it is known as a very time-consuming and labor-intensive process.

Therefore, the development of autonomous harvesting system has been actively conducted over the last decades. Several implementations have been done in the past few years for various types of crop such as strawberry, apple, tomato], kiwi], capsicum, grape, litchi, citrus, pumpkin and heavyweight crop. Most of the implementations are focusing to enhance the accuracy of harvesting system by proposing several approach and methods with different software and hardware structure architecture.

A harvesting robot is designed to pick fruits automatically under certain environmental conditions. Research on harvesting-robot-based machine vision is still in its infancy. With the development of artificial intelligence technology, 3D spatial information about the target can be obtained and processed. Stereo vision technology is a major bottleneck in harvesting robot applications, especially in crop identification, localization algorithms, error handling and small object dynamic tracking.

Here we report on fruit recognition and localization algorithms in detail by examining the following three aspects. First, the visual sensing technology; this is where stereoscopic fruit recognition and localization algorithms are expounded. Then, three techniques are used to explain how stereo vision recognizes and locates fruits under different environmental conditions. Finally, an algorithm based on 3D reconstruction is reviewed. The algorithm provides the spatial coordinates of fruit so that the robot can harvest it. Visual fault tolerance is an essential step in the successful harvesting of fruit in locations with rough terrains.

VISUAL HARVESTING ROBOT

The working environment of the visual components of a fruit harvesting robot is very complicated. The working objects are the crops, which include apple, litchi, citrus, grape, strawberry, or sweet pepper. These objects vary in size, shape, color, and texture. The background and illumination of the crops vary continuously. Machine vision-based harvesting robots should have the ability to sense and adapt to different crop types or environmental changes, collect information, detect targets, and learn autonomously. The robots should also be able to apply intelligent reasoning and engage in decision making. It is an intelligent automated machine for human-computer interaction. The robotic system should also have a network transmission function for sending the crop images to a data center or server. Agricultural robotic systems have similar structures and are composed of an autonomous mobile platform, a light multi-degree-of-freedom mechanical arm, a force feedback system with a flexible end effector, a multi-sensor machine vision system, a drive control system, an intelligent decision system, and auxiliary software and hardware.

The first task of a fruit harvesting robot is to use visual sensing to perceive and learn crop information. Its tasks include camera calibration, target recognition and positioning, target background recognition, 3D reconstruction, visual positioning-based robot behavior planning, mechanism, and vision. The system is also collaborative and uses a visual servo-control picking mechanism to perform clip-cutting operations. Changes in the illumination of complex agricultural crop environments, vibrations caused by wind or manipulators, and inaccurate positioning caused by a variety of uncertain factors can result in harvesting failure. This is the leading technical challenge in the development of visual systems for agricultural harvesting robots.

The vision system of fruit harvesting robots has several sensing capabilities (e.g., visual sensing, collaborative visual-mechanical control, visual recognition, 3D reconstruction, coordinated visual-mechanical positioning and error tolerance). Visual sensing focuses on the image-based data collection of crops. Vision-based target recognition uses various perception modalities and an accurate recognition scheme. For example, fruit imaging algorithms divide the background into a series of features. Background features, including branches, leaves, and adjacent fruits, are obstacles for mechanical operation, affecting the fruit-picking behavior of the robot. The area around the target can be labeled the target space and used for 3D reconstruction of the target. Visual-institutional coordination and error tolerance can be used to enhance robotic anti-collision precision picking guidance, picking sequence planning and the decision-making behavior of a robot

Fig. 3:Vision-based fruit picking robot system.

The robot arm of a common vision harvesting robot has at least 6 degrees of freedom (DOF); this ensures that the robot's movements are flexible. The robot has a harvest success rate of 97% and a harvest time of 31.7 ± 32.6 s. To increase the robot's flexibility, researchers studied a tomato picking robot with 7 degrees of freedom and developed a visual recognition algorithm that could identify individual fruits and bunches. The picking time of a single fruit is about 15 s and the success rate is about 70%. Researchers designed an advanced seven degree of freedom apple harvesting robot with precise positioning capability (see Figure 2B). The average positioning time of each fruit is 1.5 s, the average picking time is 6 s per fruit and the picking success rate is 84%. This robot has fast speed and can meet the needs of farmers. Adopting seven degrees of freedom improves the robot's flexibility and obstacle avoidance. However, the orchard environment was highly controlled, e.g., the clusters of fruit were removed, which reduced the complexity of the environment.

AGRICULTURAL ROBOT APPLICATIONS

Agricultural robots automate slow, repetitive and dull tasks for farmers, allowing them to focus more on improving overall production yields. Some of the most common robots in agriculture are used for:

- Harvesting and picking
- Weed control
- Autonomous mowing, pruning, seeding, spraying and thinning
- Phenotyping
- Sorting and packing
- Utility platforms

Harvesting and picking is one of the most popular robotic applications in agriculture due to the accuracy and speed that robots can achieve to improve the size of yields and reduce waste from crops being left in the field.

These applications can be difficult to automate, however. For example, a robotic system designed to pick sweet peppers encounters many obstacles. Vision systems have to determine the location and ripeness of the pepper in harsh conditions, including the presence of dust, varying light intensity, temperature swings and movement created by the wind.

But it still takes more than advanced vision systems to pick a pepper. A robotic arm has to navigate environments with just as many obstacles to delicately grasp and place a pepper. This process is very different from picking and placing a metal part on an assembly line. The agricultural robotic arm must be flexible in a dynamic environment and accurate enough not to damage the peppers as they're being picked.

Harvesting and picking robots are becoming very popular among farmers, but there are dozens of other innovative ways the agricultural industry is deploying robotic automation to improve their production yields.

Agriculture is quickly becoming an exciting high-tech industry, drawing new professionals, new companies and new investors. The technology is developing rapidly, not only advancing the production capabilities of farmers but also advancing robotics and automation technology as we know it.

At the heart of this phenomenon is the need for significantly increased production yields. The U ots are increasing production yields for farmers in various ways. From drones to autonomous tractors to robotic arms, the technology is being deployed in creative and innovative applications.

WEEDING ROBOTS

If you've ever tended a personal garden, you're well aware that weed control is both important and difficult. Commercial agricultu-rists know it, too, but on a massive scale. Even when crop rotation is possible, many large outfits rely at least somewhat on the use of herbicides. But given the facts that plants can become resistant to weed killers and consumers are increasingly averse to chemically treated food, it's hardly a perfect solution. That's why weed-management robots — including ones that incorpo-rate advanced AI to help distinguish between crops and weeds — are an attractive option.

Weeding Robot

APPLICATIONS OF ROBOTICS

• Autonomous precision seeding

Sowing seeds is the basic process to begin farming. Traditionally, farmers have been sprinkling seeds using their hands. When modern machinery came to effect, farmers used 'broadcast spreader' attacked to a tracker to sprinkle them. Although the process got simple, these attached features threw a large proportion of seeds around the field, making a complete waste of it. Thankfully, autonomous precision seeding is here to help. With a combination of robotics and geomapping, the mechanism can place the seeds exactly where it needs to be for good growth.

• Multi-talented robots for harvesting

Robots are well known for replacing humans with repetitive tasks. That is what they do at harvesting and picking. Harvesting is monotonous work that has to be done in order to reap useful food products. To relieve humans from these tedious tasks, robots are taking over the process. While planting and reaping basic food grains like wheat and barley can be done easily by robots, others like fruit and vegetable harvesting need multi-talented robots.

• Micro-spraying robots

While spraying pesticide repellent on the plants, most of the content ends up ruining the soil. Although the ground is constantly ploughed to change its texture, there is less chance for the future plantation to escape the chemicals. It is also harmful to the environment. Therefore, farmers are using micro-spraying robots to narrow down the impacts. With futuristic computer vision technology, micro-spraying robots can detect weeds and then spray a targeted drop of herbicide onto them.

• Robotic automation process (RPA)

Nursery planting is a go-to option for plan lovers. By planting the needed crops at home, we can get vegetables and fruits for our day-to-day usage. However, it is quite hectic to groom and water them on regular basis. This is where the robotic automation process gives its best. Robotic process automation takes care of all the nursery planting works including watering at regular intervals and plucking the vegetables or fruits when it is ripe.

• Robots to remove weeds

Weeds are the biggest enemies to farmers. Removing their notorious unwanted growth is both time-consuming and difficult. That is why farmers are now using robotics to counter the challenge. Autonomous robots, powered with computer vision technology, are capable of exactly identifying the weeds and yanking them out before they could spread further.

• LiDAR-powered robots to collect data

Knowing the state of the plants and the condition of the soil is as important as watering them. But a normal soil or plant analysis takes many days as the farmer has to collect samples and share them with researchers to get the results. However, the process is simplified, thanks to LiDAR-powered robots. They collect data on plants' health, physiology, and stress response and use it to improve the condition.

• Drones

Drones are not really new to farming. Since the 1980s, drones were used to capture aerial photography of fields. Even when modern drones, powered with artificial intelligence, made their debut, agriculture was one of the primary industries that took the lead to full-fledgedly implement it. Today, drones are used to capture 3D imaging, map-building, and crop monitoring.

• Autonomous agricultural robots

Agricultural autonomous robot is the most recent development that hit robotics in agriculture perspective. With various features and extraordinary technology onboard, these robots can multi-task. They can do everything starting from cloud seeding, planting seeds, doing weed control, harvesting, environmental monitoring, and soil analysis. It can take care of end-to-end agriculture processes and replace tough manual tasks with machinery.

• Robot-assisted precision irrigation

Climate change and the approaching water scarcity in the world is very real. Therefore, saving water, by all means, is at the core of farming. Unfortunately, irrigation and fertilizing processes use a lot of water and most of it goes unused. To combat this challenge, robot-assisted precision irrigation emerges as a solution to reduce water wastage by targeting specific plants.

• Sorting and packing robots

More than the actual agricultural works, sorting and packing demand a lot of human operators. In the fast-paced production space, the need for human laborers in packing is drastically increasing. Therefore, many farming companies are using sorting and packing robots to streamline the tasks at a high speed without breaking. With coordination capability and line tracking technology, these robots can fast-track the packing process.

THE FUTURE OUTLOOK

The robotics revolution is evidently happening in front of our eyes. From autonomous vehicles to automated service bots, robots are taking over a larger place in our daily lives. Agriculture is the largest manufacturing sector across the globe that is in immense need of technology. Since the global food chain is under pressure due to population growth, climate change, political pressures after migration, population drift from rural to urban regions, and the demographics of aging people, now is the right time to implement robotics applications in farming. Robotics and autonomous systems are seen as an escape from the haunting reality. It also provides a substitute solution to critical food chain transformation. Fortunately, after realizing the condition of the food crisis, governments are also coming forward to jump into the bandwagon for agricultural robotics. With central authorities in the game, robotics in agriculture and farming including dairy farming is expected to reach new heights in the coming years.

Visual Sensing Technology

The visual sensing technology of harvesting robots is designed to detect crops and fruits. A robotic servo controller collects 3D information about the environment surrounding the fruit, including geometry and 3D coordinates. The visual camera and its control system can serve as the hardware support of the visual sensing technology, which serves as a communication interface between the external environment and the robot. Images obtained by cameras are generally classified into digital images, laser images, and multi-spectral images.

Stereo-Vision Systems

Currently, two forms of stereo vision systems are mainly deployed. The first is a binocular vision system based on optical geometry. The 3D position of the target is obtained through traditional optical principles and optimization algorithms. The second is an RGB-D camera based on the time-of-flight (ToF) method, which uses an infrared sensor to obtain the depth information of the target. The ToF method is sensitive to external interference and may not work in the scene with strong light. On the other hand, the depth measurement accuracy of this method is limited by the working distance of the infrared sensor. By contrast, the optical geometry-based method is a passive measurement method, which does not rely on artificial light sources and can be used in indoor and outdoor environments. Therefore, to ensure stability in agricultural picking tasks, a binocular vision system based on optical geometry is needed. Since the principle of the RGB-D camera is simple and the system is compact, it can be used for many local tasks, such as three-dimensional reconstruction of targets at specific locations. Therefore, some brief introduction of the application of RGB-D cameras will also be briefly introduced in section 3D Reconstruction Method for Vision-based Target.

The optical geometry based stereo-vision system consists of two or more cameras separated by a fixed distance. Before the detection process, the cameras are calibrated. First, two or more images of the same target are obtained via stereo vision. The images are processed and classified to identify the target object. The 3D target is reconstructed by relating the spatial coordinates of the target to those of the robot. This relationship provides the physical parameters needed to achieve target identification and localization.

The binocular stereo vision detection technology is based on monocular vision. Early monocular vision systems used a single camera to detect one two-dimensional image of the target. Separate image analysis is performed to identify its features. With the development of computers, scholars in the 1960s began to explore theoretical research on 3D images and stereoscopic machine vision. The application of target detection in crops has also undergone a transformation from two-dimensional to 3D vision. Schertz and Brown used light information testing for fruit harvesting robots as early as the 1960s. Since the 1980s, the monocular vision was a standard component of agricultural robots. The monocular vision was used to detect the two-dimensional geometric features of crops, to identify red fruits and green leaves by detecting geometric shapes and color features. This feature detection process has a detection accuracy of about 75 per cent. As the sensing modalities and algorithms became sophisticated, researchers began to examine the role of light to obtain information about crops. These tasks often required multiple monocular cameras.

The binocular stereo vision was first used in agricultural harvesting robots to identify tomatoes, sweet peppers, and apples. Researchers installed stereo cameras in each arm of an orange harvesting robot. Stereo matching of the oranges' center-of-mass was performed to locate oranges in a 3D coordinate system. Researchers used a stereo camera to detect and locate mature apples under a canopy. The authors reported that over 89.5% of apples were successfully recognized and the errors were less than 20 mm when the measuring distance was between 400 and 1500 mm.

Researchers proposed a stereoscopic 3D vision sensing system for a palm oil-collecting robot project. The team obtained two stereo images using a mobile digital camera and used image processing algorithms for target recognition, thereby detecting palm fruit-based on image color analysis and fruit maturity-based features. The method can calculate the distance, size, and tangential position of the palm fruit. For red fruit dense images and yellow-green apple images, the fruit recognition rate was between 65 and 70%, with a ranging error of <±5%. For stereoscopic detection of a single fruit, the system needed to determine the center-of-mass coordinate of the fruit first so that the robot's fingers could grip the fruit and twist the fruiting branch. This operation of griping and twisting fruit is relatively straightforward using visual inspection. No detection of the mother branch is necessary for pinching the fruit; nevertheless, it can easily pick the fruit by using a pinching action.

To improve the picking speed of the harvesting robot, researchers studied the kiwifruit harvesting robot with four arms. Each robotic arm has a corresponding set of binocular vision to detect fruits and locate their positions in 3D space. It takes 3 s to process the complete image, and the visual recognition success rate is 76.3–89.6 per cent. The harvesting robot mechanism is flexible in design, and the four arms and four pairs of binocular vision can work cooperatively with high efficiency. However, about a quarter of the fruit still fell to the ground during the picking process. This was mainly due to interference from obstacles, which caused positioning errors or small friction between the robotic fingers.

Researchers proposed an end effector with certain versatility to clamp and cut the fruit with fault-tolerant design. The eccentric cutter is installed above the clamp so that the damage to the fruit and its body is minimal. The picking method makes stereoscopic vision detection more difficult. In addition to detecting and identifying the fruit, it also identifies and estimates the spatial position and coordinate points (also called picking points) needed to determine the clamping and cutting points of the mother branch. Two color-cameras were mounted on a six-degree-of-freedom robot that was used to locate the litchi in an unstructured environment. The litchi fruit was extracted by stereo matching two litchi images in the same scene. The recognition method is robust to changes in illumination, so 3D information can be used to recognize litchi fruits accurately. The average recognition rates of unobstructed litchi and partially occluded litchis were 98.8 and 97.5%, respectively. A binocular vision-based recognition of grapes was also investigated. The feature matching and localization capabilities of the robot to detect grapes and their picking points were shown. Due to the complexity of the terrain, the visual and stereo matching modalities are disturbed by noise, which increased the positioning error.

The challenges can be identified as follows:

- In dynamic environments, the shape of the fruit image is inconsistent with the initial shape acquired by the camera, which results in a large error in 3D positioning.
- In 3D vision technology, the calculation amount of stereo matching is great, which makes it somehow inefficient in real tasks. This is also one of the consensuses in the 3D vision field.
- With an unstructured environment, illumination and occlusion can affect the accuracy of 3D fruit detection.

LASER ACTIVE VISION TECHNOLOGY

Researchers proposed a preliminary method that used a laser-based machine vision system for automatic identification of fruits. The method uses an infrared laser ranging sensor to detect the 3D position, radius, and surface reflectivity of each spherical fruit. Combined with the proposed image analysis algorithm, the method can be applied to the citrus harvesting with <80–90% detect rate. Researchers proposed an active triangulation-based ranging system consisting of several independent laser systems. Each laser system produces a light scattering sheet that is projected onto the object. Besides, two cameras were used to solve the occlusion problem. This method integrates multiple distance measurements (obtained from these cameras) into a single image to produce very accurate depth measurements. Researchers also applied laser range finder to tomato, cucumber, and grape harvesting robots. A similar scheme for detecting mature cherries using 3D sensors was reported. The results showed that 10 of the 12 fruits were identified.

There are still problems with the use of structured light, such as complex equipment installations, unpredictable occlusions, and indirect measurements. For example, in some cases, the laser may be blocked by obstacles and cannot be projected onto the target. On the other hand, the laser may be out of focus due to long distance and result in fewer features. Furthermore, installing a laser will increase the structural complexity of the entire vision system, which is undesirable in field conditions.

MULTI-SPECTRAL IMAGING TECHNOLOGY

The multi-spectral imaging technology can image in different color spectra to see details that are invisible to the naked eye and ordinary cameras. Multi-spectral techniques divide the incident full-band or wide-band optical signal into a number of narrow-band beams, and then image the beam over the sensor. Its agricultural applications include pest and disease monitoring of fruits and crops, and growth assessment of agricultural targets. Multi-spectral technology has excellent research in the field of fruit picking. Researchers designed a system based on multi-spectral vision technology and triangulation technology to obtain the spatial position and maturity information of apple fruit through a specific optical path, which was an attempt to have great application value; researchers used multi-spectral imaging techniques to identify branches under different lighting conditions, which ensured the efficiency of path planning and safety of the operation of citrus picking robots in complex natural scenes. Researchers used different bands to image segmentation, feature extraction and classification of sweet pepper, which provided meaningful reference data for constructing obstacle maps of fruit picking robots. Researchers performed image segmentation using features extracted from multi-spectral data to provide a yield estimation method for fruit differences, and the algorithm reached high accuracy and robustness. Researchers used an automatic four-axis drone with a multi-spectral camera to capture multi-spectral images of terrain and assemble topographic mosaics to measure and evaluate green vegetation. This article provides an integrated drone solution to capture multi-spectral images with geotagging, which has important reference value in terrain and scene perception.

Researchers combined multi-spectral and drone technology and used feature extraction algorithms to detect cornfield areas damaged by wild boar. This yields a new monitoring technology that effectively acquir es terrain information and assesses the damage of crops based on its description. They obtained multispectral images of vineyards using a ten-meter-resolution satellite and a low-altitude drone platform. The system calculated three different NDVI indices and compared drone data with satellite images to analyze unbundled spectral contribution of different elements in a vineyard. The vigor map, calculated from drone images, has a higher correlation with on-site assessment compared with satellite images, revealing that the multi-spectral and drone platform can accurately complete the tasks of terrain and crop perception.

The multi-spectral technology can provide satisfactory positional and biological information for fruit harvesting tasks and is a strong auxiliary means of vision-based picking frameworks. Since multi-spectral cameras still follow the basic rules of optical imaging, they have the common disadvantages of RGB cameras, e.g., the sampling quality may be affected by changing and uneven light.

OBJECT RECOGNITION METHOD FOR HARVESTING ROBOT

Single Feature Vision Method and Improvement

The color of crops has significant and stable visual characteristics that are less dependent on the size of the image itself. In machine vision technology, an essential part of image processing is the image segmentation algorithm.

Image segmentation predicts information based on each pixel in an image. Image segmentation has two technical aspects; one is to predict only the segmentation at the class level and to mark the position for each pixel. The second is to distinguish individual objects from a set of objects.

The algorithm still needs to be improved when the vision is detected in the orchard. Researchers proposed an improved OSTU algorithm based on a fruit recognition scheme that detects and uses fruit color, e.g., red tomato and yellow persimmon would be two distinct targets. The accuracy rate of this algorithm was about 95 per cent. However, robustness is reduced when segmenting with color features, i.e., this method is sensitive to changes in the field environment, especially in the wild.

For the identification of wild crop targets, researchers proposed a Double Otsu Algorithm to segment litchi fruit orchards and achieved a correct recognition rate of <95%. To alleviate different illuminations, researchers proposed a robust image segmentation algorithm that detected illumination changes. Red litchi, purple grapes, and yellow citrus were used as examples to conduct field experiments; the detection rates were 93, 95, and 88%, respectively. When the color of the fruit is similar to the leaf color, the color characteristics are not significant, and a color-based segmentation method alone cannot be used to identify the fruit. Thus, it is necessary to combine multiple algorithms to segment the desired target.

Image shape features are mainly derived from the geometric features of the target. There are many typical feature extraction algorithms. These pre-existing algorithms often need to be improved/amended when segmenting images. The algorithm is usually unaffected by changes in illumination and is suitable for field target recognition. Researchers proposed an early single feature analysis method for citrus recognition. The method uses a circular feature to segment the citrus fruit, but the recognition rate is only 50%.

We analyzed individual images according to the geometric features of the target for the round or long-shaped fruits. The Canny operator and Hough transform algorithms are used to detect the contour of the target. This operator-transform pair was used to identify tomatoes, apples, and citrus.

Researchers combined color space and fruit geometry. The accuracy of tomato recognition in artificial greenhouses was <96%. They proposed a citrus detection method that combined a shape analysis technique and an orange detection algorithm. Experimental results show that more than 90% of the fruits in the 110 images were detected. At the same time, a controlled illumination acquisition protocol was obtained for one Flash and one non-flash image. The color-based algorithm was shown to obtain a maximum of 95% precision at a 95% recall level for FNF images, compared to 99% precision at a 69% recall for Flash-only images; this suggests the proposed FNF is effective for color-based detection algorithms.

Multi-Feature Fusion Method

Feature fusion methods combine different features to distinguish different targets. This method can improve the recognition rate of uneven illumination conditions, partially occluded surfaces, and similar background features. These algorithms are widely used for fruit recognition in conjunction with classifiers. A multi-feature integration method for vision systems was proposed to guide fruit-picking robots, clustering multiple feature detection schemes that detect color, morphological, and texture features components of the target region algorithm. The accuracy of multi-feature synthesis reached 90.15% on sunny days and 93.90% on cloudy days. A precise fruit recognition method was proposed based on a set of multiple shared features of multiple robots picking fruits in a mobile environment. The test results showed that the average accuracy of the fruit coding method reached 90 per cent. Machine vision strategies that combine strength, color, shape, and texture characteristics can also be used to identify fruits. These four features can be used to train a minimum distance classifier. The resulting experimental images were well-distinguished and had a high likelihood of accurate detection.

An artificial neural network with a RIB ratio and textured features was used to segment fruit images; for backlight illuminated image, the segmentation success rate for target in view was above 87.6%, and the bit error rate was about 13%. A fruit color recognition method was proposed based on a multi-classifier combination, which combines a support vector machine and fruit color type recognition. The experimental results showed that the average recognition rate was 86.5 per cent. Algorithms based on a combination of multiple processing and data mining techniques have been proposed to segment fruits in scenes containing different elements and to perform automated harvesting tasks in precision agricultural applications. However, multi-feature fusion methods still prone to light changes, especially in natural environments.

DEEP LEARNING METHOD

The concept of deep learning originated from the study of artificial neural networks, which have a multi-layer perceptron with multiple hidden layers. Deep learning can form more abstract high-level attribute categories or features. The high-level features are combined with low-level features to discover distributed feature representations of the data. This technology has been applied in different fields. In fruit recognition, researchers applied deep learning convolutional neural networks (CNN) to the visual techniques of agricultural robots.

APPLICATIONS OF COMPUTER VISION IN AGRICULTURE

The agricultural sector has witnessed a lot of contributions when it comes to artificial intelligence (AI) and computer vision in areas like plant health detection and monitoring, planting, weeding, harvesting, and advanced analysis of weather conditions.

Numerous smart farming use cases impact the complete food supply chain by providing useful insights about the entire farming process, facilitating real-time operational decision-making, and enhancing farming practices by introducing on-field smart sensors and devices.

How Computer Vision Is Helpful in Agriculture?

Computer Vision is a subfield of Artificial Intelligence. The emerging technology enables machines to perceive and understand the visual world like humans. Computer vision techniques, in conjunction with image acquisition through remote cameras, enable non-contact and scalable sensing solutions in agriculture.

Use cases include AI animal monitoring, visual quality control, automated inspection of quality standards, or infrastructure monitoring.

AI Technology Trends of Computer Vision

Generally, computer vision works in three basic steps:

- acquiring the image/video from a camera,
- processing the image, and
- understanding the image.

Recently, new deep learning technologies achieved great breakthroughs in the field of image recognition. Compared to traditional computer vision, modern deep learning algorithms are much more robust and allow highly accurate real-time image recognition. Hence, deep learning methods can be used to perform video analytics with the video of common surveillance cameras or webcams.

The latest trends combine edge computing with on-device Machine Learning; a method also called Edge AI. Moving AI processing from the cloud to edge devices makes it possible to run machine learning everywhere, combining IoT and AI to create scalable computer vision applications.

No-code Platform for Computer Vision Systems

To build and deploy such applications effectively, we've built a no-code computer vision platform Viso Suite that helps industry leaders across industries to deliver all their AI vision applications 10x faster and more agile. Get started and use the best computer vision capabilities out-of-the-box, and on enterprise-grade infrastructure.

Best Applications of Computer Vision in Agriculture

- Application #1: Computer Vision Systems in Livestock Farming
- Application #2: Computer Vision Systems in Poultry Farming
- Application #3: Fish Farming With Computer Vision
- Application #4: Yield Estimation With Fruit or Vegetable Counting
- Application #5: Security Monitoring for Remote Farms
- Application #6: Achieve Compliance With Animal Welfare Law
- Application #7: Drone-Based Crop Monitoring with AI

Computer Vision Systems in Livestock Farming

Food security is one of the world's biggest challenges. Livestock and poultry contribute to a large proportion (30%) of the daily protein intake through products like meat, milk, egg, and offal. Animal production is expected to increase accordingly to feed the growing human population.

As production is intensified to meet the increased demands, producers are confronted with increasing pressure to provide quality care for an increasing number of animals per management unit. This becomes even more challenging given the expected labor shortages for farm jobs in the future.

Computer Vision systems monitor animals such as cattle, sheep, pigs, or others with cameras. Neural networks are used to analyze video feeds in real-time. The advantages of computer vision systems root in the automatic, non-invasive, and low-cost animal monitoring. Vision systems allow information extraction with minimal external inferences (human adjustment of sensors, maintenance) at an affordable cost.

Computer vision is therefore needed for data collection, analysis, and decision-making in livestock farming. The insights help to improve the welfare, environment, engineering, genetics, and management of farm animals through evidence-based facility design and farm management.

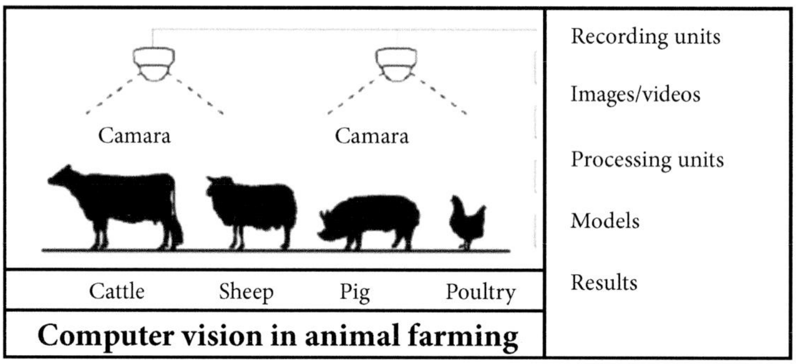

Animal monitoring systems provide continuous real-time monitoring and assist producers in management decisions. They also provide early detection and prevention of disease and production inefficiencies. AI vision is able to provide objective measures of animal behaviors and phenotypes as opposed to subjective manual observation.

Computer Vision Systems in Poultry Farming

Advanced deep learning algorithms are robust enough to be applied in poultry farming. The term "poultry" includes a range of domesticated species, including chickens, turkeys, ducks, geese, game birds, and ratites (e.g., emus and ostriches).

In poultry farms, computer vision technology aims to prevent diseases and ensure food security while enhancing overall productivity by lowering costs and providing information to increase product quality.

Today, computer vision has been widely used in poultry production systems. It includes house management automation, behavior analysis, animal welfare, disease detection, weight measurement, egg examination, and more.

Fish Farming With Computer Vision

Automatic fish detection with computer vision is an important tool in precision farming for achieving automatic fish detection. Especially, deep learning methods have shown great potential in fish species identification, counting, and behavior analysis.

Also, computer vision is rapidly developing to be used in effective intelligent feeding systems. Such systems are based on underwater image preprocessing, fish detection, wish weight and length estimation, fish behavior analysis.

Fish counting is still a rudimentary process in many fisheries. Computer vision based systems provide a cost-efficient method for counting fish with deep learning. Automatic fish counting reduces costs, helps to boost production, and increases labor availability. For example, computer vision has been used for automatic live fingerlings counting.

Related novel use cases for aquaculture enterprises involve analyzing the integrity and safety of fishing nets with deep learning machine learning technology.

Yield Estimation With Fruit or Vegetable Counting

Yield estimation is an essential preharvest practice among most large-scale farming companies. It supports decision-making for allocating essential logistics such as transportation means, labor force, supplies, and more. An overestimation leads to further costs that impact profitability; underestimation entails potential crop waste and additional costs. Yield prediction is also used to optimize cultivation practices and plant disease prevention.

Deep convolutional network algorithms are developed to facilitate the accurate yield prediction and automatic counting of fruits and vegetables on images. Modern deep learning methods provide good accuracy even with occlusion caused by leaves or branches, illumination, and object size.

Manual yield estimation with the counting of products such as fruits or vegetables is very time-consuming and expensive. Computer vision approaches can be used for the automatic counting of fruits or flowers. An example is the automatic on-tree counting and yield estimation of kiwifruit.

Security Monitoring for Remote Farms

Real-time security monitoring for remote farms is another current application of ML in smart farming. Such monitoring and notification systems are of high importance to farms. The images detected with common surveillance systems can be processed by AI algorithms to perform intrusion detection and automatically identify anomalies.

Modern methods use deep neural networks to perform accurate face recognition that is invariant to changes in illumination. This makes it possible to implement deep face recognition in multiple remote farms.

Achieve Compliance With Animal Welfare Law

Computer vision systems provide a way to automate regular on-farm monitoring to ensure compliance with animal welfare law. Deep learning algorithms and conditional logic can trigger alarms to trigger corrective actions.

Smart vision systems use AI cameras to provide objective measurements of animal welfare under field conditions. Modern methods are capable of assessing the resources provided to the animals (space, lying substrate, drinker access) and measuring the animals themselves to detect lameness, indicators of injury or disease, and abnormal behaviors. Hence, computer vision provides quantifiable data about animal welfare that can be used to ensure compliance with on-farm animal welfare.

Drone-Based Crop Monitoring

Over the last few years, drone technology has gained huge popularity because of its autonomous flying capabilities. Drones have become a significant element in precision agriculture and farming. Because of their flying abilities and cover a large distance, drones can capture huge volumes of data with a built-in camera.

Computer vision algorithms are trained with the captured footage to detect the soil conditions, analyze the aerial view of the overall agricultural land, and assess crop health information based on geo-sensing information. Therefore, the images are labeled with image annotation to create training data for algorithm training. The AI models perform object detection and semantic segmentation to recognize objects and conditions in the drone footage.

Smart systems for crop grading and sorting

AI-powered computer vision machines are in high demand in farming. The possibility to automate monotonous and time-consuming tasks such as sorting out the good and the wrong crops and finding the appropriate ones among them for shipping requirements have acted as a boon for farming.

It is becoming an integral part as the smart systems can identify the crops' longevity and find the percentage of infections in the crops, resulting in lesser crop damages. The fruits and vegetables are graded as per their quality to find which batches of the product should be shipped first and which ones may sustain for a longer time and can be sent to far destinations.

Automated pesticide spraying

In agriculture, spraying pesticides for saving the produces is a common practice. Computer vision-based drones and equipment are steadily gaining demand in the market. These machines can monitor and detect the infected crop and spray the right amount of pesticide as per the need. Such contributions have led to the workforce being healthy without being exposed to inhalation of any of the pesticides.

Phenotyping

Nowadays, the use of phenotyping to identify the crop traits for precision agriculture is immensely performed. Advanced computer vision algorithms have made phenotyping an efficient approach. The computer vision algorithms are incorporated with image processing features to eliminate unwanted crop data or information from the images while keeping only the relevant information on precise measurements.

The techniques such as depth estimation, color enhancement, identifying and segmenting the region of interest are possibilities for delivering reliable outcomes for further analysis with the acquired information. This information has significantly improved the crop breeds.

Benefits: Once again, this form of agricultural automation minimizes the manual labor required by specialists on the field to analyze each plant one by one. With the help of such technology, those working in the field can quickly acquire the data, assess the annotated information, and then take samples of plants they believe have diseases to the lab for testing.

Challenges: While computer vision models showcase great prospects for accuracy, detection, and automation, this mass of data still requires significant input by scientists. The AI must be trained to account for weather changes, plant species variations, lighting changes, and an entire plethora of diverse scientific data required to accurately assess the plants. It is still a task that requires the supervision and investment of specialists before the technology can be fully self-sufficient.

Forest information

Aerial capabilities of smart systems and drones with computer vision have made it possible to acquire data of trees that are spread out across several acres of land. Some of the possible detection based activities are tree detection, tree classification, and its types, stem analysis data, yield estimation, tree health, unused land, farming land borders, and drainage information.

Grading and sorting

Assessing features such as ripeness, color, size, and defects is a task that pomologists and other workers in agriculture carry out quite often for quality assurance of their harvest. Furthermore, they assess the needs of a particular crop, such as whether it needs more water or different antibiotics doses.

Benefits: Traditional grading and sorting is human-dependent, labor-intensive, and takes up significant time to execute for large quantities of harvested produce. More manpower is required for larger farms and manufacturers that aim to sort and grade hundreds of thousands of produce each day. That is where computer vision comes in to innovate the process via scanners equipped with image classification technology.

Challenges: Initial technology and updated prototypes yield promising results but need to be optimized for a number of specific shortcomings. Firstly, scanning of the produce via a 2D image may result in inconsistencies if the produce, for example, a tomato, has an apparent marking on the other side that was left unscanned. Additionally, the biological variation among the fruits and vegetables must be considered to yield accurate results.

Aerial imaging and scanning

Unmanned aerial vehicles (UAVs) are becoming widely utilized in recent years for tasks such as remote sensing and fire detection. There's no stopping there since aerial imaging and scanning with the help of computer vision technology aims to establish what will be known as "farming vision". The use of UAVs especially aims to be an asset for precision agriculture, otherwise referred to as aerial agriculture. Precision agriculture aims to boost eco-friendliness and efficiency while reducing the amount of traditional input required to cultivate crops such as land, water, fertilizer, herbicides, and insecticides. That can be possible through precise aerial imaging and mapping to assess each region based on the results of the imaging.

Benefits: UAVs equipped with extensively trained computer vision models are the key to smart farming by carrying out processes like vegetation classification and segmentation, crop counting, yield monitoring and prediction, crop mapping, weed identification, and disease and nutrient shortage detection. Essentially, they allow farmers and field workers to engage with crops remotely and only interfere when the data acquired from the UAVs deem necessary.

Challenges: There are still limitations to be addressed with computer vision in aerial agriculture. That includes the nuances of plant dispositions across a set period of time. For example, different plants go through different phases throughout the seasons, at different times, for a variety of intervals. The technology must be foolproof in its deep learning capabilities to assess these spatial changes without compromising on accuracy.

4

SMART FARMING AND PRECISION AGRICULTURE

SMART FARMING

Smart farming is a management concept focused on providing the agricultural industry with the infrastructure to leverage advanced technology – including big data, the cloud and the internet of things (IoT) – for tracking, monitoring, automating and analyzing operations. Also known as precision agriculture, smart farming is software-managed and sensor-monitored. Smart farming is growing in importance due to the combination of the expanding global population, the increasing demand for higher crop yield, the need to use natural resources efficiently, the rising use and sophistication of information and communication technology and the increasing need for climate-smart agriculture.

Smart farming is the best farming techniques that involve higher and advanced productivity with existing farming practices. It used to enhance production efficiency and the quality of farming products. With all this, smart farming techniques increase farmer's quality of living by minimising heavy and monotonous work.

In short, we can say that smart farming is all about reducing waste, developing productivity and full use of resources with remote sensing. Smart farming is known by precision farming too. It involves using various smart farming technologies, the internet of things (IoT) devices, big data analytics, remote sensing, artificial intelligence and robotics. It is scientifically proven that using these technologies increases profit, minimises waste and maintains the environment's quality.

Agri-tech products affordable for the farmer

Hi-tech and technologically advanced products indeed provide comfortable work with high productivity. If smart farming is not affordable for the farmers, even if they want to adopt smart farming, they won't. Most of the Indian farmers have limited income that prevents farmers from using the digital revolution in farming. Therefore, start-up agri-tech must prepare their products and services based on the farmer's budget to get the full benefits.

Smart Farming Ideas

Smart farming connected with innovation. Indian agriculture is facing huge challenges these days. There is an urgent need to accelerate agriculture by smart farming ideas. Around the last 10 years, huge growth noticed in agriculture technology investment. In the last 5 years, investment was $6.7 billion, and in the last year alone, it was $1.9 billion. These investments take place to enhance farming methods in India.

Indoor Vertical Farming

Indoor vertical farming is the procedure of growing plants in closed and controlled environments. With this method, plants mounted vertically, and it takes less land space to grow compared to traditional farming. Vertical farms don't require soil for plants to grow, and in this method, the labour force reduced too. It is the best and the first smart farming technique in India.

Farm Automation

Farm automation is the up gradation in farming machines and equipment. To accomplish this, companies are working on this. They are working on autonomous tractors, automatic watering, develop drones, robotics innovation, and seeding robots. The companies not only provide quality innovative machinery but make these machines affordable for the farmers.

Livestock Farming Technology

Livestock provides much needed products, and in our country, livestock is the most ignorant part of farming. New innovations over the 8 to 10 years created huge changes and improvements to the industry. It helps in managing and tracking livestock easily and comfortably. These technologies come in genetics, nutritional technologies, digital technology, and more.

Modern Greenhouses

Indian agriculture witnessed an increase in the greenhouse in large scale. It is urban centred and capital infused. As the market demand increases dramatically, the trend of the greenhouse is increasing in recent years. A modern greenhouse is now becoming automated control systems, tech-heavy and using LED lights for growing environment.

Blockchain

Blockchain used to resolve important issues, including food traceability, supply chain inefficiency, safety recalls, and food fraud in the food system. It creates a market for premium products with verification and transparency. This verifies transactions that are securely shared with every seller or not. It helps in creating transparency in the marketplace and food supply.

Artificial Intelligence

With the increase in digital agriculture and technologies, opportunities for the farmers increased too. Farmers can gather 24 hours of information by UAVs, satellites, and remote sensors. All these technologies can monitor the health of plants, temperature, soil condition, humidity, etc. Now farmers can get a better understanding of the ground of innovative technologies. These can help them to improve productivity.

Benefits of Smart Farming in India

Increase in Efficiency

By smart farming, farmers can increase their efficiency. With advanced technologies farmers, now farmers can produce more products in a limited time period. They get inspected fast, forecast matters before they happen and make important decisions on keeping them away.

Expansion

With the use of smart farming technologies, expansion in farming takes place. All the agriculture activities take place on time and are of good quality. The short food chain completes on time for these technologies, and everyone in the country gets proper food at an affordable price.

Proper Use of Resources

By smart farming technologies resources used fully and properly. Resources that include energy, water, and land. By the IoT farming data collected from the sensors which help to allocate an optimum amount of resources to the plants.

Cleaner process

It is a cleaner process that can save energy, water and make framing greener. These technologies evaluate the use of fertilizer and pesticides. These processes provide organic and cleaner products in comparison to traditional farming methods.

Agility

Uncertain weather changes, air quality, humidity, soil in the fields, and health of crops monitored by the smart farming technologies. That provides real time monitoring that can predict the condition of the crop. This provides professional advice in extreme weather changes that can save the crops.

Improved product quality

These help to improve product quality by using crop sensors, farm mapping, and aerial drones. Smart farming technologies create the best conditions to boost the value of nutritional products.

PRECISION AGRICULTURE

Technology has become a critical piece of business for every farmer, ag retailer, and agronomist. The increasing technology adoption rate in agriculture shouldn't be surprising to anyone. Farming is highly land- and labor-intensive. Farmers use technology to increase efficiency and manage costs.

The evolution of digital technology and the modernization of agriculture have caused new concepts (not to mention new terminology!) to emerge. These terms include precision agriculture or farming, digital farming, and smart farming. These terms are often being used interchangeably but have a subtle difference in meaning.

Popular definitions of Precision Agriculture (PA), Satellite Farming or Site Specific Crop Management (SSCM) describe the term as 'a technology-enabled approach to farming management that observes, measures, and analyzes the needs of individual fields and crops'.

According to McKinsey, the development of precision agriculture is shaped by two trends: "Big Data and Advanced Analytics Capabilities, and Robotics — aerial imagery, sensors, sophisticated local weather forecasts". In simple words farming that collects and uses data of plots for managing and optimizing the production of crops is known as Predictive farming.

Predictive farming is analogous to taking a pill to target an ailment. The solutions are highly tailored from the type of crop suitable for a plot to the use of pesticides in targeted regions only. Adopting to Precision farming reduces the production cost and wastage, as tailored needs of each plot is catered to. Precision farming is practised by adopting analytical software and use of technical equipment.

Rigorous data collection is done on soil testing, plot measurement, weather pattern analysis and crop analysis through sensor equipped devices placed along the fields. The data is calibrated to devise conclusions and based upon those results a very detailed and precise set of practices can be adopted.

As the name suggests, precision farming or precision agriculture involves highly controlled, accurate, and optimised agricultural production. It facilitates more efficient resource utilisation, better yield, and reduced environmental impact, all at the same time. An ideal example of precision agriculture practices is focussed agrochemical application with AI-aided analysis, targetting only areas that need attention instead of the blanket application.

Need for precision farming

Conventional farming practices are area-centric. There is a general set of crops cultivated throughout an area. All the farmers in that area follow the same procedures with respect to sowing, nourishing, irrigation and harvesting period. What these practices result in is: unpredictability, overuse of resources and uncontrolled waste production.

Before the use of tech in agriculture, a farmer's probability of yielding good produce was as good as tossing a coin and wishing for heads. Since farmers had no information on their farms, there was no way of learning the causes for crop loss. This practice pushed the farmers towards losses and debt. Advancements in big data analytics, IoT and accessible satellite imagery created optimism for the agriculture sector, thereby combating the issue of unpredictability.

This farming practice involves the implementation of modern information and communication technologies (ICT). Producers also incorporate a combination of devices and machinery that help capture vital field data, including:

- Remote sensors
- Automated hardware and software
- Telematics
- Drones
- Autonomous vehicles
- GPS soil sampling
- Robotics

Precision farming has been around since John Deere first introduced GPS guidance for tractors in the 1900s to enhance yield by automating steering and reducing agri-input wastage. It has evolved significantly since then, thanks to rapid technological advancements in recent years.

SMART FARMING

Smart farming focuses on the implementation of data and information technologies to utilise human labour more effectively and enhance crop quality and quantity. This approach is what drives CropIn's mission to enable agribusinesses to grow more with less.

Smart farming

Today, most farmers, especially in rural areas, rely on learnings passed down through generations and approximate estimations to carry out seeding, applying fertilisers and crop protection products, and harvesting.

Smart farming improves these processes and enhances their efficiency with the help of agritech tools and software solutions. It helps producers make more informed, data-driven decisions and achieve economic efficiency by reducing workforce requirements.

What differentiates smart farming from precision agriculture is that it does not focus on precise measurements. Instead, smart farming focuses on capturing data and interpreting them using computing technologies to make farm operations more predictable and efficient.

Smart farming employs computing technologies to transmit data from agriculture. Some of the components that ICT-based platforms use for smart farming are:

- Mobile devices, such as smartphones and tablets
- Internet of Things (IoT)
- Artificial Intelligence (AI)
- Robotics
- Unmanned Aircraft Systems (UAS)
- Sensors and Actuators
- Drones

Here is a list of operations that growers can streamline using CropIn's products.

- **Collect, process, and analyse agri-data from multiple sources for data-driven farming:** CropIn's digital agriculture platform uses satellite monitoring and weather data, information that field staff collect from farmers, and data from IoT devices, too. These data points are stored in the cloud for easy access anytime and from anywhere. The farm management software then analyses these data points and delivers actionable insights that guide decision-making processes for enterprises.
- **Automate processes to reduce human effort, cut costs, increase efficiency:** One of the primary goals of smart farming is to achieve maximum utilisation of human labour, and it does so by automating the repetitive tasks. . SmartFarm allows users to configure agronomic advisories, activity plans, and the package of practices for each crop variety and set up notifications for field agents. SmartFarm also automatically backs up the data field staff capture in the mobile app to eliminate paperwork and increase productivity.
- **Monitor operations remotely and ensure high productivity:** CropIn's SmartFarm uses satellite monitoring data from IoT devices to track plot performance continuously. It also pushes alerts to field staff to guide farmers with any possible risk of pests, diseases, or unfavourable weather conditions. Accordingly, they can cater to these issues at the earliest to ensure quality yield minimal losses.
- **Increase transparency and accountability at all levels:** Food companies and agribusinesses can also build brand reliability among their customers with SmartFarm and RootTrace. Field staff can record every farm activity in the mobile app to facilitate tracking of an end-product to, quite literally, the roots. With RootTrace, individuals can scan a given QR code to track the item's origin, safety, and quality, thus facilitating complete transparency.

PRECISION FARMING

Precision agriculture is also referred to as precision farming or precision ag. Perhaps the easiest way to understand precision agriculture is to think of it as everything that makes the practice of farming more accurate, optimized, and controlled when it comes to the growing of crops and raising of livestock. A vital component of this is the use of information technology and a wide array of items such as:

- automated hardware and software,
- autonomous vehicles,
- drones,
- GPS guidance,
- robotics,
- sensors,
- soil sampling, and
- telematics.

The critical point here is optimization. For example, instead of applying an equal amount of fertilizers over a field, precision agriculture involves measuring the within-field soil variations and adapting the fertilizer strategy accordingly. This leads to optimized fertilizer usage, saving costs, and reduced environmental impact.

Precision agriculture was initially born with John Deere's introduction of GPS guidance for tractors in the early 1990s. Their goal with this technology was to automate steering, reducing errors, and increasing crop yield by not wasting seed. Adopting this GPS guidance is so widespread that it's probably the most commonly-used example of precision agriculture today.

While precision farming principles have been around for more than twenty-five years, it's only been over the past decade that they became mainstream. This is due to technological advancements and the adoption of other, broader technologies. The adoption of mobile devices, access to high-speed internet, low-cost and reliable satellite communications – for positioning and imagery – and farm equipment optimized for precision agriculture by the manufacturer are vital technologies characterizing the trend for precision agriculture.

Here is a list of operations that growers can streamline using CropIn's products.

- Get better control over crop production and lower production risks with pixel-level intelligence: SmartRisk utilises AI and satellite imagery analysis to carry out farm plot mapping and 360-degree monitoring. Enterprises can access near-real-time analytics on crop health at plot and regional levels throughout the crop cycle. At the same time, they can receive alerts about potential risks, if any. Such continuous invigilation can help farmers determine adverse issues beforehand, thus managing crop production more efficiently.

- Optimise resource usage to decrease wastage, reduce costs, and farm sustainably: The satellite monitoring via SmartFarm can also help farmers receive advisories on the correct utilisation of water, nutrients, and crop protection products. It can help them cut down on resource wastage and unnecessary costs and aid in better management on the whole. Additionally, farming companies can get weather predictions before seeding and harvesting to determine the perfect time for both.
- Reduce the environmental impact of agricultural activities: Sustainability and precision agriculture go hand-in-hand. The implementation of digital solutions is only a way to further align with the eco-friendly narrative. Inch-perfect application of fertilisers, pesticides, and insecticides can prevent soil and water pollution due to chemical run-offs. Advisories on effective irrigation cycles and soil conservation techniques can also prevent the unnecessary depletion of these natural resources.
- Estimate yield in advance to make better decisions agronomically and post harvest: SmartRisk uses historical cropping patterns data and periodical satellite images to analyse crop health and yield. It enables farming companies to plan supply chain activities effectively, without any delay. It also helps agri-input companies develop efficient sales and marketing plans.

DIGITAL FARMING

The essence of digital farming lies in creating value from data. Digital farming means to go beyond the presence and availability of data to develop actionable intelligence and meaningful added value from such data. Digital agriculture is integrating both concepts – precision farming and smart farming.

For farmers, digital agriculture allows for the opportunity to increase their farm's production, save costs in the long-term, and eliminate risks. Many view digital agriculture as the future of the agricultural industry.

USING PRECISION AGRICULTURE DATA IN YOUR OPERATIONS

For decades, farmers have embraced leading-edge tools to help them collect and analyze farming data to increase crop yields. Soil content, moisture, nutrient trends, and weather systems down to any square foot of land are a few of the data points available to farmers with today's technology. Further, taking land and crop data and integrating that data into software used in marketing, forecasting, and production is the new "norm."

The Internet of Things (or IoT) is proving promising for realizing new levels of data and controls. It is expected to be a powerful driver that will transform farming and food into innovative webs of connected objects. There's a firm belief that IoT will be a real game-changer in agriculture and the overall food chain that drastically improves productivity and sustainability.

But to apply all of this data and take part in this transformation, the farmer needs to collect, communicate, store, retrieve, and analyze this data. The farmer needs to get inside and ahead of their decision loops. This means that they need up-to-date information and insights.

Smart Greenhouses

Smart greenhouses leverage the IoT and connected devices to create a self-regulating microclimate conducive to crop production. These controlled environments eliminate the struggles of inclement weather and predators while delivering real time insights to farmers for optimum efficiency.

Farmers using smart greenhouse crop monitoring systems can leverage insights from big data and analytics to regulate crop spraying, irrigation, lighting, temperature, humidity, and more.

AUTOMATION EQUIPMENT

Automation equipment is being used to streamline manual tasks traditionally carried out by humans, helping farmers to get back more time and resources to direct elsewhere. Some examples of automation equipment that are readily being deployed on farms include:

- Robotics. Repetitive tasks like weeding, harvesting and seeding can be carried out by robots. Utilising technologies like artificial intelligence and computer vision, robots maximise their efficiency and can work more precisely than humans.
- Smart agriculture vehicles. Using sensors, precise GPS and machine learning, smart agriculture vehicles can autonomously carry out tasks.
- Sensors. Monitor light, humidity, soil moisture and temperature. Using the collected data, farmers can accurately predict the optimal planting and fertilisation times. Sensors can also be used to map out the most efficient planting route to maximise the number of crops planted.

Control automation

Using automation control systems, farmers can adjust settings without needing to be on the farm, with the ability to relay information or tasks back to automation equipment. An irrigation management system is a key example. Using Robustel devices, farmers can set a trigger condition for irrigation to be turned off or on. This can also be used across other equipment. If an oil quality sensor shows that the oil in a system needs to be changed, an alert can be sent directly to the farmer via SMS or even email.

Irrigation Automation

In the past, irrigation was set up with a simple timer, automatically switching on at programmed intervals… but timed irrigation often can't be controlled remotely. Robustel devices can power automated irrigation solutions to allow farmers to switch on, switch off and adjust settings from a mobile device. Using sensors that are installed in areas of the irrigation system, farmers are able to view the data in real time and make decisions without needing to be on site.

Technology automation in agriculture

Automating technologies on farms can help farmers overcome a lot of factors that can make livestock and crop management a challenge. These solutions can also cut down on labour-intensive tasks, saving farmers time and money.

- Reduced physical labour. Processes like planting and spraying become automated, and in some cases can remove the need for human intervention.
- Improved decision making. Through the ongoing monitoring of conditions and collection of data, farmers can make smart and more effective decisions.
- Identify issues early. Crop diseases can be caught before they create damage and sick livestock can be identified before disease spreads.
- Reduction of cost. Through reduced resource consumption and human labour, the costs of running the farm can be better managed.

EQUIP YOUR FARM WITH IOT AGRICULTURE TECHNOLOGY

Robustel is here to help farmers find the solution for their farm. Using sensors that relay information back to control systems, farmers can adjust their systems as factors like humidity, soil wetness and temperature changes. You'll be able to access the local storage of large amounts of data to inform farm management systems.

FUTURE OF FARMING: IOT, AGRICULTURAL SENSORS, & FARMING DRONES

Smart agriculture and precision farming are taking off, but they could just be the precursors to even greater use of technology in the farming world.

The rise of blockchain technology is making its way to the IoT, and could be important in the farming sector due to its ability to provide companies with important data on crops. Farmers can use sensors to gather data about crops, which is written onto blockchain, and includes identifying factors as well as salt and sugar content and pH levels.

Insider Intelligence projects there to be nearly 12 million agricultural sensors installed globally by 2023. Additionally, tech giant IBM estimates that the average farm can generate half a million data points per day – helping farmers to improve yields and increase profits.

Given all of the potential benefits of these IoT applications in agriculture, it's understandable that farmers are increasingly turning to agricultural drones and satellites for the future of farming.

Drones allow farmers to monitor how far along crops are in their respective growth periods. Additionally, farmers can spray ailing crops via drones with substances to bring them back to life. DroneFly estimates that drones can spray fertilizer 40 to 60 times faster than doing so by hand.

ANALYZE & MANAGE FARMS WITH PRECISION AGRICULTURE TECHNOLOGY

Chetu integrates new technology into programming software precision agriculture farming solutions. We create API software that allow farmer's smart farming technologies to interface with their existing infrastructure and sensor software.

Precision Farming App

We scale precision smart farming apps and modify agriculture ERP's. Our precision AG mobile farming app development solutions ensure end to end full stack engineering design and infrastructure.

Precision Data Management

Our smart farming development uses flash memory or RAM storage then deploys data to private, hybrid or cloud-based servers to enable real-time mobile app and desktop app assessment.

AGRICULTURE DRONE SOLUTIONS

Our drone waypoint software can deploy seeds over the fields, or spray fertilizer and works in conjunction with a sensor network. Our drone software allows the drones to use built-in sensors to record data such as heat maps, crop health and more.

Soil Sensors

Our developers create software for agricultural monitoring system using digital models for soil sensors. Farmers measure pH, moisture, nutrients, temperature and more using our custom agriculture sensor technologies for precision agriculture.

Agriculture Software Integration

Our soil-mapping software, crop sensor software and drone software communicate with each other through a series of APIs and combine their data and analytics together.

Agriculture Monitoring Systems

This interoperability allows farmers to gain a far greater understanding of their crops. We provide compliance to farming and environmental regulations by monitoring much more precisely.

5

FARM, AGRICULTURAL LAND MANAGEMENT SOFTWARE

Farm management software includes websites, mobile apps, and computer programmes built for the farming sector and helps the farmer make more informed production and operational decisions:

- Aids in day to day organisation
- Makes it easier to manage workers
- Improves farm record-keeping
- Helps to comply with regulatory requirements
- Reduces modern farming's impact on the environment
- Monitors and analyses the seasons, farm activities and costs for greater efficiency and profitability

There are various farm management systems on the market, each of which include more generalised solutions for those working in multi-sector farming as well as options for single sector operations.

Budgeting and forecasting software

Designed for planning financial resources; estimating future revenues and expenses. It enables farmers, accountants, and bankers to work together by giving them real-time access to data.

Commodity management software

Used for managing commodity inventories, commodity management software tracks and reports cash flow, logistics and financials for more efficient sales and marketing.

Precision agriculture software

Precision agriculture software ensures profitability, efficiency, and sustainability by making farming activities more accurate and controlled by incorporating technology such as drones, GPS-based soil sampling and automated hardware.

Livestock and mixed farming software

Livestock and mixed farming software enable farmers to record livestock throughout their lifecycle. It can track animal inventory – including numbers and measurables – pasture rotations, grazing, and feeding plans, and provide financial tracking to record livestock sales profits.

Broadacre cropping software

Designed for crop management, broadacre cropping software monitors and optimises crop production. It helps with audits- and compliance reports, keeps the field and crop records up to date and provides a way for farmers, workers and agronomists to collaborate.

The ideal solution should be reliable, intuitive and easy to use so that you can relax and concentrate on innovating your business rather than be stuck under the paperwork. But first and foremost, it should ensure its users' data privacy. Mobile apps and offline and online capabilities of new generation cloud-based apps offer real-time data sharing. They allow farmers to keep an eye on the business wherever they are and delegate tasks to save even more valuable time.

FARM MANAGEMENT SOFTWARE FEATURES

The farm management software solutions offer various features to help farmers deliver the best results:

- **Farm mapping** – Create a digital version of your farm for an accessible overview and monitoring.
- **Paddock management** – Record and manage everything from planting and spraying to fertilisation and harvesting.
- **People and task management** -Tracking working hours, issuing work orders, including details, location, user assignment and notes.
- **Inventory management** – Keep an eye on the storage: record incoming materials and track the ones you've used up.
- **Reports** – Create more accurate farm records including the cost of production, weather information and history of farm activities and make better decisions backed by data.
- **Agricultural and financial analysis** – Learn from the mistakes and successes of previous seasons with automatically stored data. Compare management costs with planned revenue to avoid loss at the end of the season.

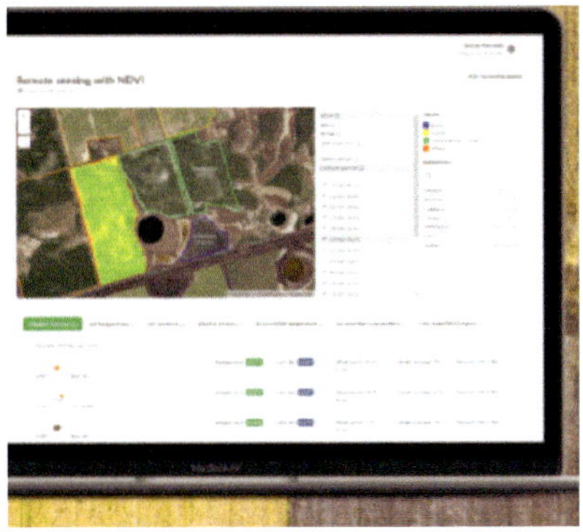

Remote Sensing With a Farm Management Software

- **Audits and regulatory compliance** – Farm management software can inform you about laws and regulations and help you run reports and prepare audits without having to fill out spreadsheets.
- **Scouting & NDVI** – The fastest way to find trouble spots on the paddock, satellite NDVI helps keep a close eye on your fields at all times.
- **Weather monitoring** – Accurate weather data on spray records helps comply with regulations and minimises the risk of litigation
- **Tracking and traceability** – This feature allows the grower to track the produce from the paddock to the buyer to increase supply chain visibility, improve quality control systems, and reduce risk.

FARM MANAGEMENT APPS

Our engineers create web and mobile apps for android and iOS devices within the agriculture industry. We scale smart farming systems and applications to access automated big data and alerts in real-time. Our secure software development integrates Open Source Farm Management Software API to configure farm management platforms to third-party vendors. We develop and modify farm supplier management software for real-time custom dashboards and visual displays to manage agriculture economics, data automation, customer contacts, and big data images using IoT agriculture devices. We also modify ERP agriculture farm management systems with modules to improve the farm calendar of operations for livestock and crops.

CROP PLANNING & LABOR MANAGEMENT SOFTWARE

We develop field operations and labor management systems to track and manage employees with scheduling, reporting, and tracking capabilities and integrate with common employee management systems. We also develop greenhouse farm management software for high-volume growers of flowers, vegetables and fruits. Our automation of watering, irrigation, planting optimization, and data analysis is enabled through farm mapping software. We also develop crop management applications, including field mapping and rotation management, crop database, yield management, and custom event-based tracking software. We create custom analytic dashboards to manage business operations and create database solutions for custom query and report creation.

FARM INVENTORY MANAGEMENT SOFTWARE

We develop agriculture inventory management systems, so the farmer can monitor materials planning, waste, traceability, and shrinkage of inventory, stockpiles, assets, and equipment remotely. Our end-to-end Farm Inventory Management Solutions provides visibility of farming equipment by the implementation of RFID systems, barcodes, serial number identification scanners, and GIS technologies to new or existing ERP platforms. We develop Farm Order Processing Management Apps to automate order processes, manage and streamline workflows, from order receipt through fulfillment solutions.

FARM EQUIPMENT MAINTENANCE SOFTWARE

Chetu's Custom Equipment Maintenance Software is the best in class. Our Farm GIS equipment maintenance software development includes mobile apps for in-field inspection and reporting capabilities and standardization of inspection protocols. EDI'S seamless transfer data to outside oversight agencies then automate backup storage to database servers for compliance. We also create software for driverless tractor technology and seamlessly integrate it with GPS satellite networks for enhanced communication and control. We also program custom m2m software solutions apps, including maintenance scheduling and tracking, Overall Equipment Efficiency (OEE) monitoring, risk and downtime management, and safety management modules and tools.

FARM ACCOUNTING SOFTWARE

Our farm accounting software control and monitor business financial operations that report, track expenses, statements, cash-flows, and balance sheets. Our custom software development for farm accounting systems and financial management includes integrations with third-party accounting software (QuickBooks, Quicken, etc.) and custom modules to calculate profit and Cost-of-Goods-Sold (COGS).

FARM REPORTING SOFTWARE

We optimize machinery performance across multiple locations, with software-for-hardware (firmware) development and machine data integration with field equipment, livestock management equipment, grower equipment, and all other farm machinery. These solutions also include integration with data warehousing solutions, to capture and store information for analysis and systematic record-keeping.

AGRICULTURAL LAND MANAGEMENT SOFTWARE

Mapping

We program precision agriculture software for site-specific spatial data mapping through GIS / GPS software technology and integrate with cost analysis applications to create seed yield forecasts and field optimized planting

Planting Technology

We develop custom prescriptive planting software technology to support acreage and crop maximization by identifying historical crop yields and weather patterns, seed performance, and soil analysis with site topography maps, creating custom forecast reporting capabilities for specific crop yields

Harvesting Software

We develop harvesting software modules to manage inventory, loads, dispersals, contracts and track yields from each field from one robust management platform

Decision Support System

We program custom Decision Support System (DSS) applications to enhance rate seeding and yield applications, allowing multiple scenario testing and enhanced predictive analysis before planting

Crop Management

We develop site-specific crop management and production cycle platforms with integrated, secured access portals for stakeholders including growers, agronomists, service providers, GIS specialists, and others to access real-time field information from any location

3D Field Design

We program custom 3D field design applications to integrate with topography mapping software for detailed visualization of land shapes / slopes, wet / dry zones, drainage areas, and levy locations, to plan waste management / irrigation layouts, and perform robust field yield analysis during planning

IRRIGATION

The advantages of smart irrigation are far-reaching. By monitoring soil moisture levels, a smart water irrigation system allows farmers to automate their irrigation processes and reduce water use. In addition to more efficient consumption of resources, other benefits include:

- Cost savings due to minimized water waste
- Reduced human efforts
- A unified view of soil characteristics, including moisture and nutrient contents
- Smart notifications in case of abnormalities
- Better long-term landscape health
- IoT ecosystem for smart irrigation

To achieve these advantages, smart water irrigation systems make extensive use of IoT sensors. These sensors, placed in the field, send real-time data to a central gateway that then automatically switches on a water pump whenever moisture or temperature values are outside the predetermined range.

Wireless low-power networks like LoRa are used to empower IoT sensors and make it possible for information to flow in real time to and from the central gateway. The entire smart irrigation system can be managed by an end user through a custom cloud-based platform or mobile application.

Types of IoT sensors for smart irrigation of farmlands

Depending on the type of data to be captured, soil, weather, and plant IoT sensors can be used in a smart irrigation solution.

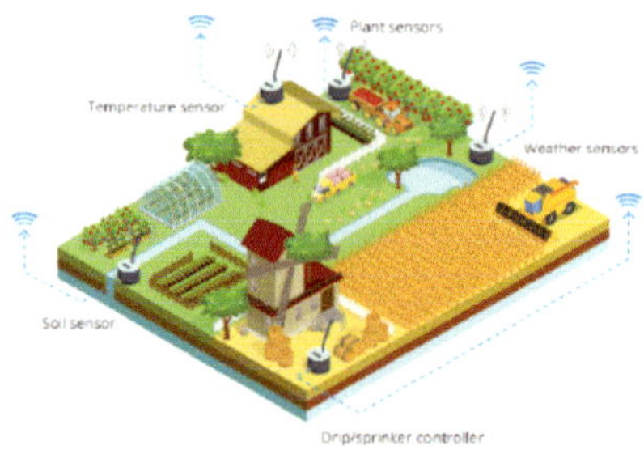

Soil sensors

Soil-based sensors gather relevant data about volumetric water content, salinity, electrical conductivity, and other crucial parameters. Located at key points across the field, these sensors send data to a smart water irrigation system to help farmers gain quick insights into the soil's state and predict irrigation needs.

Weather sensors

Also called evapotranspiration (ET) sensors, weather sensors measure ultra-local environmental conditions like water evaporation from the soil surface and plant transpiration. Combined with data provided by a GIS-based solution, these sensors can help generate more accurate water predictions.

Plant sensors

Plant-mounted sensors are still new but show great promise in gauging a plant's water status. Attached to a plant's stalk or fruits, a sensor can track minute changes like swelling or shrinking and alert farmers to insufficient water content or yield-reducing tendencies in the field.

EXTENDING THE CAPABILITIES OF A SMART IRRIGATION SOLUTION

An IoT-based irrigation system can be further enhanced with functionality that uses advanced software. You can customize your smart water irrigation solution and integrate third-party APIs to tap into additional valuable data and inform your decision-making on irrigation of farmlands.

Weather monitoring

One way to do that is to incorporate satellite data and weather reports from weather stations to better schedule your irrigation activities. Knowing that rain is forecasted, the system can wait and automatically recalculate the amount of required water based on actual precipitation received.

Location technology

Precision mapping also helps agribusinesses drive smart farming by optimizing costs and gaining deep insights. Reusable maps that bring together data from sensors and images from satellites and drones enable farmers to keep an eye on critical land characteristics.

Artificial intelligence

AI is becoming the bedrock of automation in farming. Apart from its use in novel but still rarely used agricultural robotics systems, AI helps to automate simple tasks like labelling data, building reports, and sending notifications. Such small steps to automation through AI are becoming easier to implement while their impact on agribusinesses is critical and can be seen in just a few months.

HOW TECHNOLOGY REINFORCES EACH TYPE OF IRRIGATION SYSTEM

Based on how water is distributed throughout the field, you can choose from different types of irrigation systems that can be enhanced with smart irrigation software. The most common are flood, sprinkler, center pivot, drip, and micro-irrigation systems. Let's see how technological solutions for smart irrigation can improve the efficiency of each type.

Sprinkler irrigation

In a sprinkler irrigation system, water is pumped through pipes and then distributed via high-pressure overhead sprinklers. These sprinklers can be set in a central location in the field or can be located on a moving platform.

Role of software: Thermal and acoustic rain sensors recognize rainfall and measure its intensity to schedule the next irrigation after rain stops. A smart irrigation system analyzes data and calculates the water budget for the next month. Sprinklers get automated notifications to prevent extensive water use and overwatering due to rain.

Center pivot irrigation

This is the most popular form of sprinkler irrigation and is also known as water-wheel and circle irrigation. A typical center pivot system consists of a long irrigating pipeline attached to a central tower and moves slowly over the field in a circular pattern, irrigating plants with sprayers.

Role of software: The system that controls circle irrigation sprinklers obtains data insights from in-field sensors to adjust the water stream or angle of flow. This helps to reach plants that are far from the water source and save those nearest from overwatering. By analyzing weather data and soil moisture, the system plans irrigation and calculates potential yield and harvest times.

DRIP IRRIGATION

In this type of irrigation, water is distributed directly to the roots of plants through pipes with small openings called drippers. This allows farmers to significantly reduce evaporation and runoff.

Role of software: For this type of irrigation, the main challenge is the visibility of the watering process. The system notifies the user through an app about starting and finishing irrigation. It also measures soil parameters before and after irrigation.

Micro-irrigation

Micro-irrigation is a low-pressure, low-volume system that offers precise control over watering. The system applies water directly to the plant's roots, improving irrigation efficiency and ensuring uniform distribution.

Role of software: The system can plan the exact dosage for each plant as the amount of water is precisely controlled. AI algorithms can be applied to recognize plants and adjust watering appropriately.

No matter what type of irrigation system you choose, equipping it with powerful IoT sensors will help you gain data-driven, actionable insights and stay on top of your irrigation needs.

REMOTE IRRIGATION CONTROL

Monitor irrigation from phone or web

Farmers have spent countless hours and tanks of gas at all times of the day and night driving back and forth to fields to check irrigation systems. With this remote control technology, a farmer can now save time and fuel with the ability to remotely monitor and control irrigation systems from their phone or computer.

This technology controls multiple aspects of the system remotely: turn on, turn off, change direction, set a timer for the system to start, control the end-gun sprinkler with precision by degree, check the position of the system, set the stop for a precise angle, and even speed control variable rate irrigation on certain systems. All systems enabled with this technology are available on a singular dashboard and even connected with some soil moisture sensor systems transmitting real-time data from the field.

This technology is available for both center pivot and drip irrigation systems.

Efficiency

No official studies have been done to approximate the efficiency of remote control technology for irrigation systems. Any increase in efficiency is highly subjective by field and producer. The ability to stop the system at precise angles, control the end-gun by degree, and set automatic timers to start and stop the system take limitations on timing out of irrigation management (e.g. if the farmer is unable to arrive at a system at the precise moment it needs to be stopped).

FARM MANAGEMENT SOFTWARE SYSTEMS

Farmers often need to respond to challenging physical environments determined by climate, soil type and water availability. Well-managed farms can better protect themselves against threats posed by harsh conditions, and that's where farm management software can lend a helping hand. Let's take a look at a few examples of the best farm management software systems:

Figured

Farm financial management software designed explicitly by two farmers and accountants. Access livestock and crop information and financial data all in one place. The features include budgeting and planning, livestock, crop and milk production tracking, multi-farm management, and stock valuation calculation.

Eka

This cloud-native CTRM/ETRM solution allows businesses to manage the entire trading life cycle with advanced analytics to adapt to rapidly changing market dynamics and offers. The software involves a streamlined sustainability solution, including pre built sustainability reports, for monitoring the strategic use of energy and resources.

John Deere

John Deere's leading precision agriculture software helps connect people, machines, technology, and insights. Collecting agronomic and machine operation data from sensors in the machines and the field – and information like weather and commodity pricing from external sources – optimises farm activities such as planting, tillage, and harvest and helps improve farm management.

Agriwebb

AgriWebb is an agriculture software company aiming to transform the livestock industry by helping farmers with record-keeping, compliance and productivity. The software harnesses simplified on-farm data collection and analysis for improved management efficiency and allows transparency, profitability and sustainability across the supply chain.

Conservis

Suitable for row crop and permanent crop platforms alike, Conservis is a cloud-based comprehensive farm management software system that connects data across the operation for better business management. It provides traceability through features like real-time record-keeping, planning and budgeting and allows connecting field data to finances for better risk management, cost control and improved profitability.

eAgronom

eAgronom is a cloud-based farm management software designed for grain growers to improve their record-keeping, organize their tasks and plans, and analyse their seasons. It starts with the paddock map – the whole farm and the crops grown on each paddock. Crop planner allows creating the rotation for up to five years into the future with a few clicks. Tracking the tasks in eAgronom results in clear and easy record keeping. The inventory feature helps growers stay up to date with products and avoid unexpectedly running out of stock. eAgronom also has a mobile app that works offline. When the season's over, eAgronom puts growers' data to use. The analysis screen allows comparing yields, task schedules, fertility, and profit. The agronomists have their own accounts in eAgronom, from which they'll be able to access your farm and work alongside the grower.

Scouting With a Farm Management Software

OTHER SOFTWARE IN AGRICULTURE

In addition to farm management systems, there are other helpful software to improve and streamline agriculture operations. Here are a few examples:

Smart Irrigation Software

These systems, used in conjunction with other agriculture software, optimise water usage throughout the crop lifecycle by helping businesses make informed decisions based on the collected data.

Feed Mill Software

Nutrient values vary according to animals and their age, requiring the change in feed throughout their lifecycle. The software can solve this otherwise time-consuming task and calculate the suitable formulation within no time.

Farm Safety Apps

Farm safety apps can help farmers identify hazards, assess and control risks, review control measures, record and report incidents, capture health and safety training records, record team meetings, and store emergency plans.

AGRONOMY

Soil Sampling

We develop custom agronomy management software including soil sampling analysis applications and automate results processing through integrations with outside labs and databases including the USDA Natural Resources Conservation Service's Field Office Technical Guide (FOTG)

GIS Mapping

We integrate GIS mapping with historical datasets for soil type and characteristics identification, and spatial analysis information processing for efficient crop management

Soil Collection

We engineer soil collection automation software and program routing capabilities to save and evaluate samples by grids / zones, depth, size, GPS coordinates, subsoil type, and more

Subsurface Drainage Systems

We program wet soil management software to design agricultural subsurface drainage systems with integrated drainage level calculators

Soil Color Index Mapping

We program crop or soil color index mapping software and spectral algorithms to isolate soil components

Soil Fertility Automation

We develop custom soil fertility automation software for nutrient recommendations and calculators that integrate to crop forecasting modules with multiple file export options and wireless data transfer capabilities

Satellite And Aerial Imagery

Integration of crop / soil mapping software with satellite / aerial imagery applications to identify land configurations including irrigation patterns, organics, sand, clay subsoil, and variations from crop residue

Seeding Software

We program variable rate seeding software to automate planting rates with soil productivity potential and integrate with outside databases for predictive analysis

OPERATIONS MANAGEMENT AND REPORTING

Unmanned Aerial Vehicles

Software-for-hardware (firmware) development and device integrations for aerial imagery cameras, Unmanned Aerial Vehicles (UAV), and remote sensors

API Integrations

We develop mobile applications with 3rd party API integrations for irrigation management capabilities from tablets and smartphones

Spatial Data

Spatial data centralization through database warehousing platforms with custom business intelligence analytics applications to support query and reporting capabilities for soil samples, weather information, and GIS map storage capabilities across multiple fields and crops, plus migrations from existing systems

Chemical Application Tracking

Chemical application tracking modules development to maintain compliance with regulatory and environmental safety standards for pesticides, fertilizers, and other chemicals

Data Warehouses

We implement data warehouses to store crop yield information, global predictive weather modeling information, and design custom reporting capabilities to pull site, crop, and yield specific measurements allowing analysis over time, seasonal comparisons, and historical benchmarking

We integrate with 3rd party software to pull information from field naming standardization databases to ensure yearly operational consistency

AGRICULTURE DRONE SOFTWARE

Agriculture Drone Spraying Software

Drone Agriculture Mapping Software

We provide smart farming technology using ground-based electromagnetic, optical and acoustic sensors. Our agricultural drone flight software has 3D mapping capabilities. Farmers generate real time analytics of seed patterns, metrics or 3D computational models using our drone flight software dev.

Aerial Crop Analysis

We provide farmers usage of aerial crop imaging and aerial sensor technology to scout and detect irrigation issues, soil variance, and crop infestation. Our precision agriculture software provides drone data findings from crop scouting drones for autonomous field management.

Smart Farming System Solutions

Our software configures drones and allows them to perform aerial spraying over a field: either fertilizer, pesticide or another regulation-approved liquid. This aerial spraying maximizes crop absorption compared to ground spraying, and lessens the effect of ground absorption.

UAV Data Processing Software

Our Unmanned Aerial Vehicles (UAVs) provides orthomosaic stitching for agriculture mapping. Our precision agriculture software designs implements cloud-based data management solutions to cloud servers for smart farming drone data analysis, storage and reporting using 3D modeling API interfaces for accurate drone image processing.

LIVESTOCK MANAGEMENT WITH UAVS

Farmers track and evaluate livestock and agriculture remotely using custom mobile app data. Chetu's custom software allows GPS tagging, thermal recognition and enables AI farming solutions for predators and health using drone technology.

AG Drone Development Platforms

Our agricultural drones programming enables sensors and software assess and monitor the health of ag assets remotely. Our software controls the distribution and placement of these seed pods, and interfaces the sensor network with the drones themselves through APIs.

Agricultural Drone Analysis Technology

We integrate drone software platforms with Normalized Difference Vegetation Index (NDVI) indicators for crop health analyses.

PRECISION AGRICULTURE DRONE SOFTWARE

The software developers provide custom and open source development of AG drone software to generate a 3D map, detect crop health, distribute seeds or spray crops and monitor livestock remotely.

6

HARDWARE AND COMPONENTS FOR AGRICULTURE MONITORING SOLUTIONS

Agriculture has transformed in the last few decades from small to medium farming operations to highly industrialized, commercial farming that is concentrated among a few large corporations. However, as various Internet of Things (IoT) technologies mature beyond the R&D phase and go into general production, costs for everything from drones/UAVs to sensors will continually decrease, making connected agriculture more accessible to smaller farms and third world countries.

With this agricultural transformation, farming operations are increasingly a highly mechanized and computer-driven operation. This allows corporations to treat agriculture like manufacturing in the sense that measurements, data, and control is very important to manage costs, maximize yields, and boost profits. This shift in managing agricultural operations will bring various benefits to farming and livestock management, including enhanced crop quality and quantity, improved use of resources and farm equipment, real-time monitoring of farms, animals and machines, automated irrigation systems, fertilizer spraying and pest control.

The general term, AgriTech, represents the use of technology in agriculture, horticulture, and aquaculture for purposes of improving yield, efficiency, and profitability. The commercial agriculture industry is rapidly becoming one of the most IoT data-driven markets. With the emergence of M2M, IoT, and advanced data analytics technologies, data is becoming available that was previously uncollectible. The application of various AgriTech analytics tools and methodologies, such as predictive analytics will provide substantial enhancements to agriculture operations.

IoT in Agriculture (IoTAg) represents a more specific use of technology wherein agricultural planning and operations becomes connected in ways previously impossible if it were not for advances in sensors, communications, data analytics and other areas.

Virtually every aspect of agriculture that can be automated, digitally planned, and managed will benefit from IoT technologies and solutions.

Accordingly, we see IoTAg fundamentally transforming the way agricultural operations and farms are managed, which will bring various benefits to farming, including enhanced crop quality and quantity; improved use of resources and farm equipment; real-time monitoring of farms, animals, and machines; and automated irrigation systems, fertilizer spraying, and pest control.

The implementation of IoTAg is intended to facilitate greater agricultural efficiency and effectiveness. Essentially, IoTAg solutions, coupled with artificial intelligence and a few other supporting technologies, enable smart agriculture. IoTAg solutions provide many intelligent agriculture benefits such as increase of yields, monitoring crops, automating operations, and reducing waste.

IoT technologies allow farmers and ranchers to enhance productivity. For example, if part of the irrigation system malfunctions, sensors can provide alerts, allowing the problem to be addressed in a timely fashion. These technologies also allow agricultural staff to view operational conditions from anywhere and make changes with real-time solutions.

There has been a paradigm shift in the use of technology in agriculture sector in the last decade, however, farmers from the developing countries are still lacking in modern tools usage primarily because of cost, unavailability of products in local market and lack of education. Human force is generally used for all the steps of crop growth starting from sowing, irrigation, monitoring the growth, pest control till harvesting. In order to cope up with the demand of food ingredients and enhance the economic returns, it is necessary to adopt autonomous/semiautonomous systems which can help and assist humans to have more productivity. There have been many success stories around the globe where robots are used to pick citrus fruits in United States, harvest tomatoes and pick up strawberries in China, picking dates from palm tree in Kingdom of Saudi Arabia, sowing rice by robots in Japan and irrigating crops autonomously in many parts of the world.

The agriculture IoT market is expected to grow at a CAGR of 15.2% from 2020 to reach $32.75 billion by 2027. The growth of the agriculture IoT market is driven by the factors such as positive encouragement from local governments and development organizations for the adoption of modernized farming techniques. Furthermore, the rising population and subsequently growing demand for high-quality food products and consistently shrinking arable land are the other prominent factors promoting the growth of the agriculture IoT market. Besides, the factors such as increasing awareness about AI and big data analytics, consistent proliferation of IoT devices across different end-use sectors, and enhanced end-user experience are also contributing to the growth of this market.

On the basis of application, precision farming segment is estimated to account for the largest share of the overall agriculture IoT market. With consistently growing population which is expected to reach almost 10 billion by 2050 as stated by United

Nation Organization, the demand for high-quality food products is expected to pose one of the biggest challenges to human race. In order to counter such issues and increase the productivity of limited arable land, governments are encouraging farmers to adopt IoT enabled precision farming techniques. Thus, the precision farming segment dominated the overall agriculture IoT market. However, the smart greenhouse segment is expected to register maximum growth rate over the coming years. Regulation of controlled environment can help boost the productivity within limited resources in greenhouse agriculture. Integration of IoT in greenhouse farming can help to achieve high production rate of good quality food. Hence, the smart greenhouse segment is expected to grow with a healthy growth rate over the forecast period.

An in-depth analysis of the geographical scenario of this market provides detailed qualitative and quantitative insights about the five regions including North America, Europe, Asia Pacific, Latin America, and the Middle East and Africa. In 2020, North America is estimated to command the largest share of the global agriculture IoT market, followed by Europe and Asia Pacific. The large share of this region is mainly attributed to its open-minded approach towards smart agriculture and high technology adoption rate, presence of key players & start-ups, and established infrastructure. However, the factors such as high spending power, participation of young population in modernize farming, and government initiatives supporting agricultural digitalization is helping Asia Pacific to register the fastest growth in the global agriculture IoT market.

THE CURRENT AGRICULTURE INDUSTRY CHALLENGES

Agriculture is rightfully considered to be one of the most resource and labor intensive industries. The challenges farmers face today include, but aren't limited to the following:

Regular equipment maintenance

Agriculture as an industry is heavily dependent on machinery. Maintenance operations, even scheduled regularly, consume time and impact the budget; but nevertheless, fail to eliminate the unpredictability factor. Once a piece of equipment accidentally goes out of order, it usually leads to unexpected downtimes.

Correct water estimates

Growing plants need water, but the amounts of it differ depending on soil humidity levels. To measure these levels, farmers have to go to the field and take regular manual tests – alternatively, they could use smart sensing technology, which is by far, more accurate, convenient, and time efficient.

Eliminating water waste and overhead expenses

Failing to collect accurate soil humidity information may result in the underwatering or overwatering of the plants. Poorly watered plants are dry and frail, but overwatering creates water waste and involves unpredicted water expenses.

Estimating correct planting times

Each plant has its own optimal planting time depending on a range of environmental factors. However, it is often difficult to correctly estimate this time without accurate data.

Measuring soil temperature and moisture levels

Soil temperature and moisture levels are key metrics farmers need to collect to estimate the state of crops and take appropriate action. Unfortunately, it's usually impossible to measure them correctly without IoT agriculture monitoring systems.

Pest control

Successful pest control involving detecting pests, their location, activity and behavior patterns is another challenge farmers have to face. Understandably, this challenge is also quite difficult to meet without IoT based pest control systems.

Smart agriculture monitoring solutions

IBM predicts, the use of IoT will enable farmers to increase production rates by 70% by the end of 2050, so, all in all, the future looks optimistic. One way or the other, IoT has a lot to offer in terms of alleviating the pains farmers regularly face.

Agritech is a thriving industry, and, as of today, an extensive range of smart farming systems enables farmers to meet their daily challenges. Planting, watering, crop gathering and pest control – agriculture field monitoring collects a range of metrics farmers can act on to manage these tasks effectively.

Below are some examples of smart agriculture monitoring solutions and how they work:

SOIL CONDITION MONITORING

Soil condition is an important indicator helping farmers decide on the optimal planting, and crop gathering time. With IoT sensors performing soil condition monitoring, farmers get instantly alerted of soil moisture and salinity. Other metrics include soil temperature and air temperature: estimating them correctly enables farmers to plan watering times and know when to expect pests.

Soil condition monitoring requires a combination of hardware and software systems to operate in real-time and alert users on any significant changes.

Soil condition monitoring

An example of such a solution is CropX – an ag-tech platform for agriculture remote monitoring. It uses smart agriculture sensors to collect data, and a cloud infrastructure for data processing and storage to deliver information in a readable format to a user's computer or smartphone screen.

WEATHER MONITORING

Weather monitoring in agriculture is one of the most frequent application fields for IoT. In crop farming, yields are heavily dependent on the environment, which is inherently volatile. Weather monitoring solutions located directly in the field (such as the ones used by weather stations), alert farmers on changing weather conditions – temperature, precipitation, humidity, sun radiation, and wind speed.

Weather monitoring

Weather monitoring platforms like Pycno, allMETEO, and Smart Element are vivid examples of how the application of smart sensing technology in agriculture helps deliver effective weather notifications directly to farmers' laptops and smartphones, enabling them to immediately take action.

GREENHOUSE AUTOMATION SYSTEMS

A fragile and sensitive greenhouse ecosystem requires incessant maintenance and control. Smart agriculture solutions for greenhouse automation like Growlink, Farmapp, and GreenIQ illustrate the application of remote sensing in agriculture. They help maintain optimal microclimate conditions and manage lighting, humidity, $CO2$ and temperature levels. Instant alerts and increased management capabilities maximize the efficiency of greenhouse farming.

CROP MONITORING SYSTEMS

As crops grow and ripen, so many things can go wrong: diseases, infestations with pests, or adverse environmental conditions can potentially cause irrevocable harm before farmers even notice. Applied in crop monitoring, smart sensing technology collects metrics about the state of the crops (temperature, humidity, health indicators) and enables farmers take timely measures should anything go wrong.

Moreover, systems like Semios and Arable help detect when the crop is ripe, enabling farmers to plan exact harvesting times.

DIGITAL PEST MANAGEMENT

Pest infestations are some of the pains crop farmers face on a regular basis. Knowing when pests arrive can be challenging, but also pinpointing their activity and location is normally impossible without making frequent trips to the field. Smart agriculture monitoring systems tackle these problems; moreover, they also help allocate the exact amount of chemicals needed to eliminate pests in each particular case.

IoT pest detection systems like Strider count insects and determine their locations in real-time using an insect camera and sensors for crop pest detection placed directly in the field. Ag-tech companies like Fieldin and DTN offer similar solutions for IoT-based pest control.

LIVESTOCK MONITORING SYSTEMS

Apart from crop and weather monitoring, agriculture monitoring solutions are also gaining wider application in livestock farming. By combining sophisticated IoT hardware such as wearables based on smart-sensing technology with state-of-the art IoT software, ag-tech solutions like Cowlar help guard and protect livestock.

SCR is another company specializing in agriculture remote monitoring using cow neck collars to track cow health, location, and activity. Remote sensing in agriculture, combined with advanced analytical software delivers insights on cow nutrition and on the health of the entire herd.

END-TO-END FARM MANAGEMENT SYSTEMS

From greenhouses to grazing fields, the entire farm area can accommodate smart agriculture sensors acting as important data collecting points for a powerful, all-encompassing farm management system. Surely, such systems should leverage advanced data analytics software and integrate seamlessly with accounting and procurement databases to deliver insights and fully unveil their analytical potential.

Cropio and Farmlogs are examples of companies offering end-to-end agritech solutions for remote farm management based on IoT agriculture monitoring.

HARDWARE ARCHITECTURE OF FARMING ROBOT

Mechanical System

Pinon belt mechanism for farming robot structure.

Our proposed farming robot, is in terms of mechanical structure, is primarily based on threedimensional Cartesian coordinate system namely X axis, Y axis and Z axis. This means that our mechanical structure can move in these coordinates. The movement of X and Y axis is governed by "Pinion Belt" mechanism which fixes the belt with the track and allow motors to move along with the construction along the track. Pinion belt mechanism is very useful for linear motion application as of ours. The Pinion Belt mechanism, as shown in figure, was chosen because of its flexibility to adjust the belt length, lower cost of the belt and the lower cost of maintenance. The drawback of this mechanism is that it is not very precise, however, the precision is enough to operate the system when used with the stepper motor with encoder.

For Z axis the "Lead Screw" mechanism is used for up and down movements which needs to be a lot more precise than Pinon belt. A lead screw has a thread optimized for reducing friction and is used frequently in many applications. Precise positioning and nonoverhauling can be obtained with a shorter lead, however, the main downside of lead screws is that they are mechanical inefficiency converting 20% to 80% of the applied torque into linear thrust and rest of the energy is wasted in the form of heat. Furthermore, in our proposed scheme, the lead screw mechanism is made selflocking as well which will prevent any accident or damage to the crops. The lead screw mechanism is shown in figure below. Figure (a) shows the motor of the linear stage which is connected to lead screw with a coupling. Motor is used to drive the lead screw. Figure (b) is the nut of the lead screw which is used to mount the tool and is also responsible for its vertical movement.

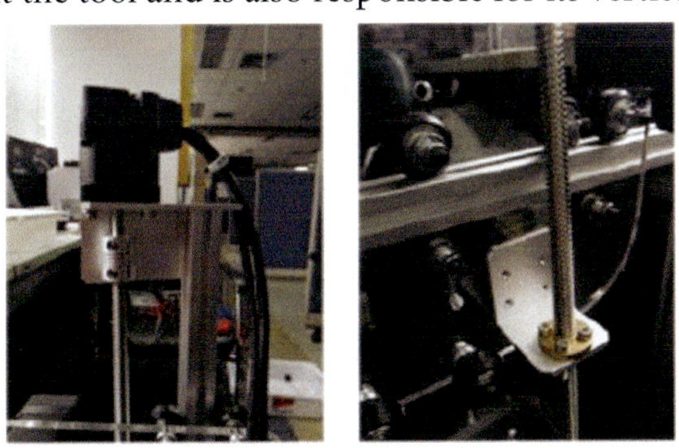

a **b**

Lead screw mechanism: (a) Top and (b) Bottom view.

For the system to operate in three independent axis, four step motors are employed, two of which are used for the motion in X axis, one is on the top of the gantry for movement in the Y axis and another is installed at the top to drive the lead screw, perpendicular to the gantry. Each motor is driven by a separate Sanyo Denki motor driver SANMOTION PBM0007846D using PWM technique with 12 volts power supply.

External view of stepper motor

Each stepper motor (as shown in Figure) has 500 steps per revolution which is set at the motor driver. 32000 steps per second speed is used in all axis with 1000 steps per second acceleration and deceleration for smooth movement.

An Arduino Mega 2560 board as is responsible for the simultaneous control of all four motor drivers. This facilitates the universal tool mount to move in any desired location with any tools including the watering tool and the seeder. In addition, the Arduino board is used to control two solenoid valves for the watering tool and a vacuum pump for the seeder, and these tools are controlled through relays.

For control calibrations, limiting switches are installed at both ends of each axis, allowing the system to calculate the farm space. All switches are connected

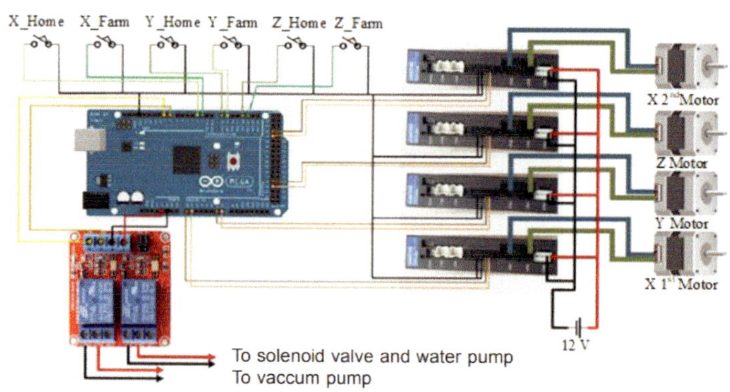

Connection among various peripherals

to the Arduino board. The system consists of many wires such as the step signal, direction signal, limiting switch signal and power supply wiring. All wiring can be concluded in a wiring diagram. Plastic cable holders are used to hold all the wiring needed. The holders move with the mechanism, but they are rigid enough to hold their own weight with some additional support.

TOOLS SYSTEM

der tool and a watering tool. The universal tool mount is fixed at the lower end of the Z axis and can be moved around the farming area according to the control command received.

The seeder and watering tools are placed on a tool holder plate fixed on the aluminum profile frame. All tool parts are made by 3D printing method, using ABS plastic with 20 percent density.

The watering tool and the seeder tool design is shown in figure. It can be firmly attached to the universal tool mount according to the three magnets as presented in Fig. (b). The watering tool and the seeder tool designs are basically

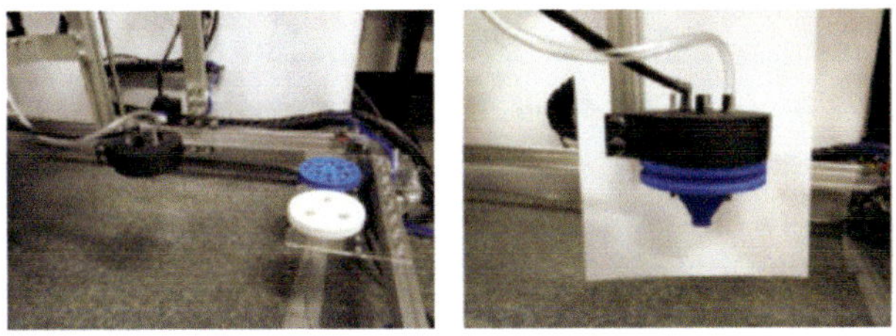

a b

(a) Watering and seeder tool; (b) Watering tool mount with universal tool mount.

the same; consists of three holes for mounting magnets, another three holes made to be compatible with the universal tool mount which has the project tubes, and nine holes reserved for the purpose of sensor in the future. The water source for the watering tool and the vacuum pump for the seeder tool are connected to the tool function using water tube and vacuum tube respectively and is controlled using Arduino Mega 2560 Microcontroller through active high relays.

The control system for the tool function is shown as figure. From designs, the watering tool is recommended using with 3/16" tube with connector, which could release satisfying amount of water and could prevent leakage. The smallest size of seed that the seeder can deliver without slipping into the tube is about 2 mm diameter.

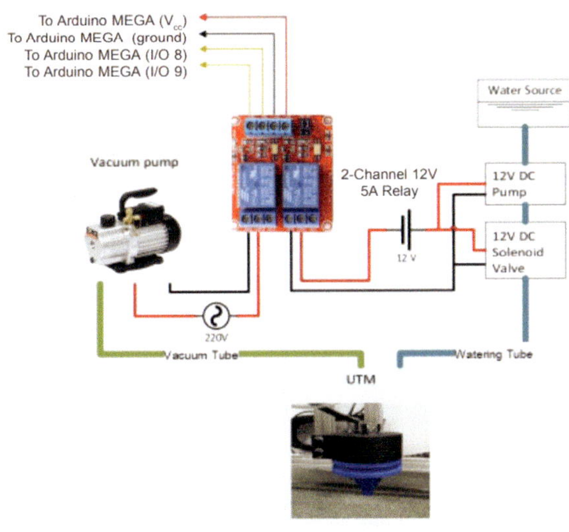

Overall structure of tool control

SOFTWARE ARCHITECTURE OF FARMING ROBOT

Overall Software Architecture

There are three main components in it. The first one is the development of graphical user interface (GUI) which is used to control and govern the farming robot. User is not directly connected with the farming robot but with the help of cloud service. These services are provided by Network Platform for Internet of Everything (NETPIE). So the instructions are routed from the user to the farming robot via NETPIE. The farming robot has two main components. The first one is the Raspberry Pi which is working as a server. Clients can connect to IP of this Raspberry-pi in order to communicate. The raspberry-pi sends commands to the Arduino (the second part of farming robot) which is responsible for motions of the farming robot's motors. Arduino board is facilitating the function of simultaneous control of all four linear bipolar motor drivers. The universal tool mount has the ability to move to any desired location carrying the watering and seeding tools. Furthermore, the Arduino is also responsible to control the solenoid valves for the watering tool and a vacuum pump for the seeder.

Initially, the program builds a connection between Raspberry-pi and Arduino. These two board communicated with each other via serial communication. Python serial library is added first and then Arduino's port is defined which establishes a connection. Then, the Raspberry-pi (which is acting as a server) has an IPv4 address. The program assigns an IP address to the server and the port 8000. The server has a provision to establish connection with the number of clients simultaneously.

Once connection is established, client is ready to send the signals to the Raspberry-pi over the local network. Finally, using a loop, we make our server to receive commands from clients iteratively otherwise the server can take only one command and get close. Unicode character encodings "UTF-8" is used for encoding the inputs.

THE BENEFITS OF USING IOT MONITORING SOLUTIONS IN AGRICULTURE

So, how does monitoring and recording data improve agriculture? The range of agriculture remote monitoring applications is quite extensive, and so is their combined effect on livestock and crop farming.

All in all, the use of IoT monitoring solutions accounts for the following:

Maximized productivity

Agriculture crop monitoring using IoT and taking timely measures to eliminate the usual threats increases crop yields. In livestock farming, the use of IoT monitoring also makes for maximized productivity.

Improved quality

IoT monitoring systems help maintain optimal conditions to ensure better crop quality. For example, weather monitoring in agriculture helps estimate the exact supply of water, chemicals, and nutrients needed to grow high-quality crop yields. The farming products grown using IoT monitoring systems are also more capable of meeting market specifications than other products.

Reduced need for pesticides

Not only are pesticides toxic, their use also entails expenses. Smart pest monitoring systems significantly reduce the need for pesticides, the expenses involved, and the dangerous impact of chemicals on the environment and human health.

Predictability and control

Driven by realtime agriculture monitoring, data analytics predict the optimal harvest dates and ensure the security of supply contracts. The control farmers gain over time-to-market helps make farm processes more manageable.

Higher sales price

Obviously, greener, healthier products grown using the latest ag-technologies will have higher sales prices, and will ultimately bring more revenue.

Future-casting

By collecting and processing data retrieved using smart agriculture monitoring, farmers can predict the future state of soils and environments, and plan for next year's crops. Thus, predictive analytics enable them to make calculated farm-management decisions and plan for the years ahead.

7

SOFTWARE COMPONENTS FOR MANAGING SOIL, SEEDS, IRRIGATION AND FERTILIZERS

Software is a generic term for organized collections of computer data and instructions, often broken into two major categories: system software that provides the basic non-task-specific functions of the computer, and application software which is used by users to accomplish specific tasks.

SOIL TESTING FOR SMART FARMING

Traditional methods used for determining soil quality are expensive and time-consuming. Therefore, soil testing laboratories need to digitally transform their operations to support smart farming. The Internet of Things (IoT) has been a trend in the agriculture sector to determine weather conditions, check soil quality, and analyze crop health. IoT-based smart sensors such as humidity and temperature sensors and cameras accumulate a huge amount of data. IoT-based smart farming techniques help monitor the field in real-time, improving the entire agriculture system. IoT-enabled technologies and wireless communication techniques can help in all stages of farming from sowing and harvesting to packaging and transportation.

Smart Farming

Smart farming leverages emerging technologies such as IoT and cloud computing to determine soil quality based on various factors. Various IoT sensors are used to collect environmental and machine metrics for decision-making on improving soil quality. Big data applications also play an important role in smart farming. The IoT sensors are capable of capturing huge volumes of data for further analysis. Big data provides insights into farming operations, helps in real-time decision-making, and enhances business processes. The agriculture sector is likely to benefit significantly from the implementation of IoT applications.

Benefits of Smart Farming

- Enables collection of massive amounts of data through smart sensors.
- Reduces cost and crop wastage by analyzing potential damages at early stages.
- Increases production efficiency by automating processes.
- Increases productivity and enhances quality.
- Supports informed-decision making.
- Helps save resources such as water, energy, and land.

SOIL TESTING & EMERGING TECHNOLOGIES

Soil testing laboratories play a crucial role in determining the quality of the soil. However, the turnaround time is slow using traditional testing methods due to manual processes. Furthermore, test results are reported on paper and sent manually. Due to delayed test results, farmers fail to make timely decisions on sowing crops.

Technologies such as IoT have improved soil testing using smart sensors. IoT helps connect the sensors to the cloud, enabling farmers to determine the soil parameters such as pH, moisture content, and NPK content in real-time. IoT enables all laboratory instruments to connect with different networks, informatics tools, smart devices, and also facilitates data interoperability. It helps in the digital transformation of a laboratory, increasing its efficiency. Soil testing laboratories must consider deploying a cloud-based informatics tool to support IoT for seamless functioning.

Digitization: Role of a Laboratory Information Management System

IoT sensors can measure soil temperature, NPK, moisture content, and soil oxygen levels among others. The metrics collected by these sensors are then transmitted to a central database for storage and downstream analysis. These metrics help farmers to make data-driven decisions, thus making precision farming a reality.

An Agriculture Laboratory Information Management System (LIMS) is essential for soil-testing laboratories for improved and digitized workflows. All analytical instruments and IoT-enabled devices in a laboratory can be integrated with a LIMS. A LIMS solution for agriculture & farming stores all data collected from different devices and analytical instruments in a common database, thus allowing easy access. It also eliminates human errors and assures data integrity as the data from the devices is automatically transferred to a LIMS. It provides real-time access to test results. A LIMS helps generate custom Certificates of Analysis (CoAs) that contain the details of samples, tests, and test results. A LIMS for soil testing labs helps comply with quality control standards to support the accuracy and validity of test results. A LIMS solution for agriculture & farming helps minimize the test turnaround time without compromising the quality of test results.

Technologies such as cloud computing, IoT, big data, and informatics solutions are revolutionizing the agricultural sector. IoT helps in precision farming and helps farmers make smart decisions before they sow crops. It provides farmers with necessary information on soil nutrients, climate conditions, soil moisture content, and soil health. The traditional soil testing methods do not provide highly accurate results and are also costly and time-consuming. Soil testing laboratories need to be digitized for ensuring the quality of test results and reducing turnaround time. A LIMS solution for agriculture & farming is a perfect data management and automation system that goes hand in hand with IoT technology. It helps laboratories automate all laboratory processes from seed to sale and meet regulatory compliance such as ISO/IEC 17025:2017, GLP, and 21 CFR Part 11.

SOIL AND CROP HEALTH MONITORING SYSTEM

The type of soil and nutrition of soil plays an important factor in the type of crop is grown and the quality of the crop. Due to increasing, deforestation soil quality is degrading and it's hard to determine the quality of the soil.

A German-based tech start-up PEAT has developed an AI-based application called Plantix that can identify the nutrient deficiencies in soil including plant pests and diseases by which farmers can also get an idea to use fertilizer which helps to improve harvest quality. This app uses image recognition-based technology. The farmer can capture images of plants using smartphones. We can also see soil restoration techniques with tips and other solutions through short videos on this application.

Similarly, Trace Genomics is another machine learning-based company that helps farmers to do a soil analysis to farmers. Such type of app helps farmers to monitor soil and crop's health conditions and produce healthy crops with a higher level of productivity.

Farmers are now using sensors to monitor particular sectors of the field, enabling them to react quickly to changes in the land and crops. The use of smaller and less-complex sensors is making quick response possible. This allows farmers to turn soil sensor readings, weather, and historical crop data into actionable perception by seeing the bigger picture.

Soil is never consistent across a field and this inconsistency is often amplified at the sensor level. Multiple sensors can statistically improve the accuracy and track the active changes, which are variable across a field. A wet area in the spring may become dry later as the crop grows and uses up the water. Sensor-based measurements are providing more specifics, such as moisture levels, fertilizer effectiveness, and plant reaction to variable conditions, including temperature and light. These sensor measurements permit farmers to take action when a field condition, such as low water levels, produces a stress reaction.

Soil moisture sensors estimate the volume of water content based on the dielectric constant of the soil. The dielectric constant indicates the soil's capability to transmit electricity. As the water content of the soil increases, the dielectric constant of the soil increases, because the dielectric constant of water is much larger than the other soil components, including air. Therefore, the measurement of the dielectric constant gives a predictable assessment of water content.

Most soil sensors are single-point sensors. They take a measurement at a single location. A single point sensor can measure soil moisture and temperature, or soil moisture, temperature and salinity. The sensor can be completely buried in the soil. Some sensors measure volumetric water content for the length of the sensor.

Profiling probes measure soil moisture across a vertical soil profile, covering a range of one foot to four feet. Soil profiling probes comprise of multiple single-point sensors contained in an elongated enclosure. Some, like the GroPoint Profile include modular segments which form a single antenna for constant measurement across its entire length.

Fig. 2: Edaphic Scientific soil moisture and temperature sensor

It's important to measure soil moisture at multiple depths in order to optimize irrigation, as it indicates the penetration of water all through the root zone. The main benefit of using a soil profiling probe is the reduction of costs of installing multiple single-point sensors, and the requirement to dig a large hole to bury them at the proper depths.

Profiling probes are typically manufactured as parallel pairs of rings along a probe, and are normally installed in plastic or PVC access tubes where the electric field between the sensor and the soil must pass through the tube. Profiling probes, which don't entail an access tube, will deliver greater accuracy. The GroPoint Profile and Hydrascout HSTi are examples.

Fig. 3: GroPoint Profiling Probe

Sensors that measure the volumetric water content are typically referred to as soil moisture sensors. Soil volumetric water content sensors measure the water content of soil. These sensors can be used to estimate the amount of stored water in a profile or how much irrigation is required to reach a desired amount of water in the soil. These sensors can be used for quick measurements or installed for long-term measurements.

Volumetric water content is a measure of the amount of water held in a soil expressed as a percentage of the total mixture, and is often called simply "soil moisture". The amount of water that can be stored by a soil and its availability to plants both depend on soil type.

Soil sensors are permanently buried and are left for continuous long-term monitoring if they are connected to a data logger, or wireless remote telemetry, or on-demand monitoring using a handheld reader. They can remain buried indefinitely, depending on the durability of the sensor and the cable. Portable soil sensors provide the user an instantaneous reading of soil moisture in a battery-powered, self-contained unit that can be carried anywhere. Readings are shown on an integrated display, or on the user's smartphone, which communicates with the sensor unit through Bluetooth or WiFi.

The amount of moisture or soil water is important to know because soil water performs as a carrier of food nutrients needed for plant growth. Soil water is a nutrient by itself. The yield of a crop is most often influenced by the amount of water available rather than the deficit of other food nutrients. Water regulates soil temperatures. Microorganisms require water for their metabolic activities, while helping in the chemical and biological activities of the soil.

SEEDING AND IRRIGATION SYSTEM

Cropping is important and tedious activity for any farmer, and for large scale this activity is so lengthy also it needs more workers. Thus agriculture machines were developed to simplify the human efforts. In manual method of seed planting, we get results such as low seed placement, less spacing efficiencies and serious back ache for the farmer. This also limited the size of field that can be planted. Hence for achieving best performance from a seed planter, the above limits should be optimized. Thus we need to make proper design of the agriculture machine and also selection of the components is also required on the machine to suit the needs of crops. The agriculture is the backbone of India. And for sustainable growth of India development of agriculture plays vital role. The farmer has to spend his more time in planting. But time available is less for him. Thus it requires more man power to complete the task within stipulated time which is costlier. Also more wastage happens during manual planting. Hence there is need of developing such a machine which will help the farmer to reduce his efforts while planting. This process of using machines is called as mechanization. Along with mechanization automation also helps to increase the efficiency of the process.

The robotic system is an electromechanical (conveys a sense that it has agency of its own) and artificial agent which is steered by DC motor which has four wheels. The farm is cultivated by the machine, depending on the crop considering particular rows & specific columns. The infrared sensor detects the obstacles in the path and it also senses turning position of vehicle at end of land. The seed block can be detected and solved using water pressure. The machine can be controlled remotely and solar panel is used to charge DC battery. Assembly language is used in programming the microcontrollers. The microcontroller is used to control and monitor the process of system motion of vehicle with the help of DC motor. As agriculture is extensively supported by technical means like seeding, mowing or harvesting machines, it is widely considered to be a field with a high potential for robotic application as it is a small step from these semi automatically operated machines to fully autonomous robots in both greenhouse and open field applications. Robots are available on all development levels from experimental to market-ready in several agricultural applications but most of them are in research, where institutes have made progress to extend the existing agricultural machines to robotic systems. Most of the robots considered in this publication are developed for harvesting. Seeding is not yet as important since there are already good tractor based seeding systems. In horticulture there are significantly less robotic applications as in agriculture.

THE AUTOMATED SEED SOWING TECHNOLOGY

The Automated seed sowing technology is a method design in order to reduce the human efforts as it requires less amount of manmade labour and can be handle efficiently without a skilled operator. Seeding manually requires lots of time, therefore this technology develops which eradicated much amount of time with proper efficiency, less time consuming, accuracy in sowing seed at specific distance.

It works on simple mechanism, a battery operated D.C. motor is used transmits the rotary motion to the shaft with the help of chain drive, and there is another connection of sprocket and chain to the rotary motion. When the farmer puts seeds into the hopper. As the seed rotates, seed drops in the seed pipe, which is connected to the furrow opener for the seeding; there is furrow closer for covering the seeds by soil.

COMPONENTS OF SEED SOWING MACHINE

Basic Components

A Seed sowing machine is constructed using the following components:

D.C. Motors

It is used in the model to drive the front wheels which further drives the distributer.

Hoppers

It stores the seeds to be sown in the soil. Higher the capacity less the need to refill the hopper during process.

Seed Distributor

It consists of fluted rollers which are driven by rear wheel with the help of belt and pulley.

Cultivator

The work of the cultivator is to tilt the soil to the required depth so the distributer mechanism can sow the seed.

Belt and pulley drive

For transmission of power from motor to wheels belt and pulley drive is used in the machine and also to drive seed distributer. The material used in fabrication of the machine is mild steel grade (MS). MS are used for manufacturing of truck-trailers, cranes, ship building, boilers, agricultural equipment and in many more fabrication and engineering industries.

Electrical Components

For automating the seed sowing operation the following electrical components are used:

- Single channel relay
- Step Down module
- Wi-Fi ESP8266 microcontroller
- Jumper wires
- Battery (12 V)

Specifications of the Machine

The specifications of the machine is presented in table 1.

Table 1 Specifications of the seed sowing machine

Length	610 mm
Width	356 mm
Height	254 mm
Power transmission	Belt and Pulley
Hopper limit	1 kg medium size seeds
Number of distributer rollers	2
Distribution Mechanism	Fluted roller
Distributer driver	Rear wheel

SOFTWARE FOR SEED PRODUCTION MANAGEMENT

Brought together in a single, same ERP business, all the configurable functions allow you to feed inventory projections and cover all production processes :

- Production management (production plan, provision of basic seeds, contracting, technical monitoring)

- Management of manufacturing plants and stocks (seeds and consumables)
- Laboratory Management
- Sales management (order, delivery, batch invoicing and prices)

A Cloud-based multilingual offer, agreo SEEDS safeguards and consolidates all automatically processed and analyzed technical, agronomic and administrative information – for the entire profession and stakeholders of all sizes – in France and abroad (cooperatives, trades, breeders, multipliers, production organizations, research stations or official control services). This new version announces real savings in time and productivity, allowing, for example, a threefold reduction in the process of producer payments.

Fig. 4: Software for seed production management

In the age of Mobile First, this solution adapts to different environments and work constraints (PC and mobile media) and also integrates mobile applications dedicated to both technical monitoring (agreo mobile inspection) and the barcode management of production orders at the factory.

SMAG currently has about 40 key accounts from the industry among its users, some of whom are among the ten largest multinational seed companies in the world (Groupe Limagrain, Sakata). This year, the company recorded a 30% increase in its Seeds software business with the signing of new major accounts such as Deleplanque, DSV and Excelience, which is a new plant co-built by the Noriap, Cap Seine and Uneal French cooperatives to consolidate their Seeds business.

SEED COATING EQUIPMENT AND METHODS

The seed treatment and coating materials described in Section 2 provides an extensive list of potential ingredients. The next step in the seed coating process is when selected ingredients are applied with appropriate equipment to produce the final coated product. The selection of seed coating equipment and coating method is determined primarily by the dosage of actives, liquids and solid components applied per unit of seed. There are three major types of seed coating equipment used today: dry coating, rotary pan and pelleting pan. This coating equipment used singly or in some cases in tandem is paired with five coating methods: dry powder, seed dressing, film coating, encrusting and pelleting. The overall goal of all coating equipment and methods is to achieve good application uniformity and adherence. Processes should not cause mechanical injury to seeds during coating.

OPEN-SOURCE SOFTWARE SYSTEM

Besides the seed imaging module, the SeedGerm software system also contains a light-weight data management module and a MLbased analysis module (Fig. 5a). An image acquired by the imaging module is firstly saved on the SeedGerm hardware's local storage. Then, the image is checked for its size and clarity; if the size and clarity are greater than a predefined threshold value, it will be transferred to a gateway computer via wired (ethernet) or wireless network connections. This synchronisation task is carried out on an hourly basis, between many SeedGerm devices and the gateway computer, where images from different devices are collated in folders named after their associated experiments defined through the seed imaging module. Between the gateway machine and onsite storage (e.g. a dedicated workstation or high-performance computing infrastructure, HPC), data synchronisation tasks are normally accomplished overnight, when onsite network traffic is less busy. The data management module is administered by either crontab scheduling on LINUX (Debian 9.0 onwards) or Bash scripting on WINDOWS (WINDOWS 7 onwards), which have been described in our previous work.

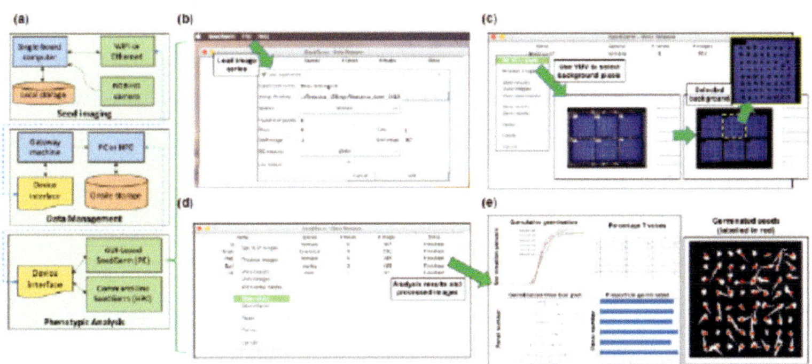

Fig. 5: (a) Three software modules designed for automated seed imaging, data management and machine-learning based phenotypic analysis. (b) The input parameters that can be set by users before automated phenotypic analysis. (c) The background pixels in germination panels (i.e. filter paper) identified using user-configured YUV colourspace ranges (where Y stands for brightness, U and V for colour components). (d) After processing a number of experiments, germination and morphological traits are quantified and ready to be downloaded. (e) Cumulative germination curves, uniform and Gmax plots produced to score seed quality and seed vigour together with dynamic seed masks recording entire germination procedures for different genotypes in a germination box

For different end-users, automated phenotypic analysis can be conducted either centrally on a HPC or in a distributed manner on a workstation. Each experiment generates a time-lapse image series, which is uploaded to the onsite storage progressively during the experiment. Then, users can use the ML-based phenotypic analysis module (i.e. analysis software) to either analyse these images through a command-line interface on HPC clusters, or tailored GUI-based software on a normal workstation computer. Both approaches output similar analysis results, including the quantification of germination- and establishment-related traits in CSV files, as well as a sequence of processed images (e.g. dynamic seed masks and panel segmentation images) in PNG format.

GUI-based analysis software

For the ML-based phenotypic analysis module, the workflows for both GUI and command-line approaches are fundamentally identical. We therefore use the more accessible graphic user interface (GUI) software to introduce the analysis procedure, which has been designed to execute on either WINDOWS (i.e. the .exe executable, WINDOWS 10 tested) or Mac OS (i.e. the .app file, v.10 onwards). The analysis software packages can be downloaded from our GitHub repository. The initial GUI contains an empty window with a menu bar, and experiments can be added via the "Add experiment" window (Fig. 5b), through which users can enter a given experiment's name, select an image series for processing, and choose a crop species such as Brassica, maize, pepper, tomato, or cereals. New plant species can be trained and added to the software through the Modules directory, an approach that is independent of the core analysis algorithm. Users need to briefly define the germination experiment associated with the selected image series, including the number of panels in a given SeedGerm device, and the number of Rows and Columns of seeds in each panel. In particular, users can define the Start and End image IDs to initiate and terminate the phenotypic analysis, because the background in early images can be over-saturated due to excess water soaked by the filter paper, whereas late images can contain too many overgrown seedling and roots (e.g. images between the 4th and 167th image are analysed in Fig. 5b). Default values of the Start and End images are the first and last image of the selected series.

In order to deal with varied image quality and features caused by lighting, crop species, and different establishment phases, a number of ML-based algorithms have been implemented in the software. Users can select the ML technique from the 'BG remover' dropdown to remove the background pixels, which includes U-Net, Gaussian mixture model (GMM), and stochastic gradient descent (SGD) (Bottou & Bousquet, 2008), which are explained in the Core analysis algorithm section. After an experiment is added, users are required to set YUV colour-space ranges (Y stands for the brightness, U and V for colour components) to delineate the background (i.e. filter paper) in the first image of the selected series (Fig. 5c). By adjusting the sliding bars in the 'Set YUV ranges' window, backgrounds are mostly retained, representing different types of filter paper used in diverse experiments. After defining YUV values, users can click the 'Process images' item to start the phenotypic analysis (Fig. 5c). Similar to our previous work, the analysis software has also employed parallel computing to process multiple experiments simultaneously, and up to 12 image series have been analysed at a time on an average computer (Intel Core i5, 8GB RAM) and over 120 series on a HPC (Fig. 5d). This implementation has enabled a multi-threading analysis running on HPC clusters for greater throughput.

Finally, when the analysis is completed, germination traits (e.g. T25, T50, T75, Gmax, and germination timing curves for each panel), morphological traits (e.g. area, width and length, extent, convex area, and circularity for each seed), and a range of processed images (showing the germination procedure and labelling individual seeds) are produced (Fig. 5e). Users can click 'View results' on the shortcut menu to display the analysis outputs, as well as download a range of processed images (Video S1) and the analysis results in CSV files, containing phenotypic analysis at the image (overall results), the panel (i.e. a given genotype), and the seed levels.

Core analysis algorithm

The core analysis algorithm for SeedGerm includes three key parts: an ML-based background remover, feature extraction and germination detection, and traits measurement (Fig. 6). To establish a more general algorithm to analyse different types of seeds robustly, we have used a mixture of deep learning (DL, i.e. UNet) and supervised ML (i.e. GMM and SGD) to divide background (filter paper) and foreground (seeds) pixels. For example, after users set the YUV values to retain background pixels, the selected BG remover is trained based on features of the background (e.g. RGB, contrast, intensity values) in the image (Fig. 6a). Then, the YUV values are applied to representative images across the image series (i.e. images at the beginning, middle, and end of the series) to segment background pixels, which allows the ML model to learn background features at different establishment stages during a given experiment, without overfitting the classifier for a specific crop species or a particular experimental setting. Finally, the trained classifier (i.e. the background remover) is applied to each image in the series, producing background masks excluding any seed for each germination panel (Fig. 6b).

After producing background masks, they are inverted so that only seed-related objects can be retained. SGD has been chosen as the main learning algorithm for germination scoring and was used for our routine germination experiments because it performs well when the seed–background contrast is high when seeds can be clearly delineated from the surrounding background pixels. Unlike SGD, the GMM model is used for image series with low seed–background contrast when seeds are slightly out-of-focus. It is slower, but it is more robust when the background is complex (e.g. roots from different seeds are crossing). For images with acceptable quality but under changeable lighting conditions, we tend to use U-Net, a recent convolutional neural network (CNN) for semantic segmentation, which excels the Fully Convolutional Network by adding skip connections and extra upsampling layers to provide both local and global information. The implementation of U-Net is exploratory, with the aim of using deep learning techniques to improve analysis for unseen datasets (e.g. treated seeds and new plant species).

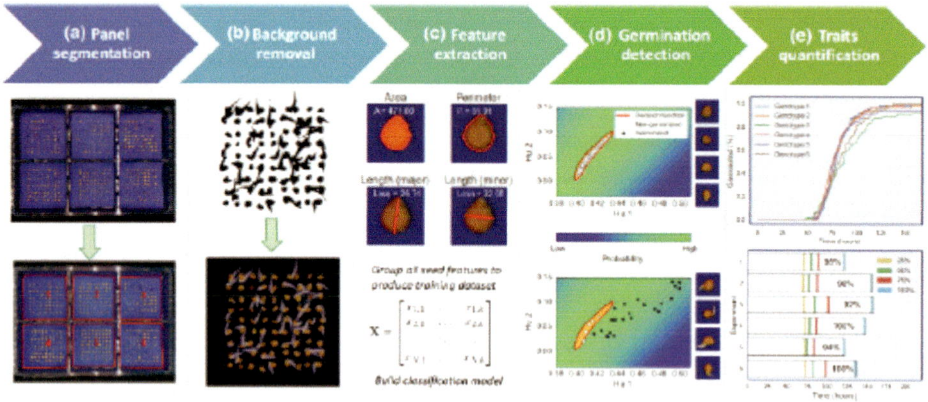

Fig. 6: Analysis software for automated seed germination scoring and phenotypic analysis. (a) Location of individual experiments in a given germination box, which is identified from the first image by panel segmentation using the YUV colour-space ranges (where Y stands for brightness, U and V for colour components). (b) Background removal models trained using labelled background (filter paper) and foreground (seeds) pixels, which are subsequently applied to retain seed-related objects. (c) Descriptive statistical moments (Hu moments) are used to measure seed morphological features for each seed in a given germination panel. Features across the image series are combined to form a set of training vectors. (d) A novelty detection model (one-class support vector machine (SVM)) is trained based on the training vectors and is then applied to the time series to classify the germination status for each seed, germinated (closed circles) and non-germinated (open circles). (e) Germination scoring and morphological traits for all seeds are collated to produce cumulative germination curves and germination timing plots.

For feature extraction and germination detection (Fig. 6c,d), we applied descriptive statistical moments to describe a given seed's area and its centroid position, which are invariant to the scale and rotation changes of seeds due to imbibition in early germination stages. Features such as minor axis length (seed width), major axis length (seed length), length and width ratio, perimeter, delta of Hu moments (i.e. difference of a seed's Hu moments between two consecutive images), delta of seed area, and delta of seed length and width, have been computed to monitor a seed's morphological changes. The aforementioned features from the pre-germination stage are extracted from the first 20% of images in the series and are then combined to form a training matrix (Fig. 6c) to train a classification model with the assumption that the label for all the seeds is non-germinated.

The detection model used is called novelty detection, which is a one-class support vector machine (SVM) established on the training matrix and is then used to determine the germination status of each seed in the image series.

Based on the training data, a decision function is generated by the model to enclose pre-germination feature vectors (shown as white circles enclosed by the red-coloured contour in the embedding p-dimensional space shown in Fig. 6d); then, as the germination experiment progresses, feature vectors are recomputed. When a seed begins to germinate, its feature vector should gradually leave the boundary of the initial observation region (i.e. abnormal with a given confidence in the germination assessment – black circles outside the red-coloured contour in Fig. 6d). The seed's probability of germination will increase as well. The novelty detection model scores germination for all detected seeds, resulting in cumulative germination rates for each seed lot in a given germination panel (Fig. 6e).

Since the novelty detection model is reinitialised and retrained for each experiment using the first 20% pre-germination images of the selected image series as training data, the detection model is dynamic and hence the risk of overfitting is low.

THE SEEDGERM SOFTWARE DESIGN

There has been a growing need for the standardisation of plant phenotyping in recent years, resulting in the ISA-Tab format, Minimal Information About Plant Phenotyping Experiments (MIAPPE), and ontology approaches to enable comparative phenomics research. Much previous work in seed phenotyping has employed bespoke data collection processes and data formats, meaning external researchers and laboratories have been limited to the use and support of these methods. When designing SeedGerm's software system, we therefore chose to standardise the collection of image and sensor datasets following the ontological suggestions.

Additionally, to calibrate images acquired by different SeedGerm devices, users were required to enter metadata to define their experiments, including experiment ID, genotype, biological replicates, treatment, and experiment duration; imaging intervals, image resolution, white balance, exposure mode, and shutter speed were then controlled automatically by the imaging module to largely standardise the data collection.

To increase the scalability of the phenotypic analysis, we chose to implement our algorithms in PYTHON instead of MATLAB, as previously reported (Jahnke et al., 2016; Elmasry *et al.*, 2019). The reasons are that PYTHON is easy to understand, cross-platform, and self-contained, and is supported by a wide range of open-source libraries such as Scikit- Image, OpenCV (Howse,), Scikit-Learn, and Keras/TensorFlow.

Publicly available development kits have enabled us to extend and upgrade our software relatively easily. For example, new crop species and traits can be added to the core analysis algorithm through new modules, where guideline seed morphological features can be predefined. Also, as we implemented the software based on a modular software design, modules developed for one species can be shared by other functions in analysis and parallel computing.

Recently, deep learning has become a powerful technique used by some seed germination analysis software, in which it was applied to extract features, segment seeds, and classify germination status. Although DL is presently relatively easy to implement through PYTHON, the reasons we chose a combined CV and ML approach are as follows.

First, DL requires a very large amount of training datasets to perform better than supervised ML and CVbased methods – for features that need to be engineered frequently such as varied seed germination experiments, DL might not be suitable.

Second, we normally need to build a dedicated DL model of each species; hence, it is time-consuming and ineffective to employ DL techniques to analyse a large number of crop species.

Third, DL is likely to overfit for particular experiment settings and becomes problematic when conditions are changed. To allow our solution to be adopted by a broader research community that has varied experimental settings, we chose supervised GMM, SGD and novelty detection learning techniques based on generalised feature selection.

More importantly, by designing the ML models to reinitialise and retrain with background features at different establishment stages for each experimental setting, the learning models embedded in SeedGerm are dynamic and can be updated for each experiment, enabling us to avoid overfitting the learning models for a specific crop species or a particular experiment.

By employing CV algorithms, SeedGerm can also measure cumulative germination rates and seed morphologies such as size, width and length, extent and circularity to assess seed quality and seedling vigour, from germination to seedling. For example, we have measured imbibition using the change of seed size, radical protrusion based on seed major : minor ratio, and germination speed through seed extent. If new biological questions are proposed, new traits and features could be designed jointly by biologists and computer scientists, instead of blindly relying on DL techniques. Because the SeedGerm software can be easily extended and accessed, we believe it is scalable and easy to access.

APPLICATIONS OF THE SEEDGERM SYSTEM

Our work has demonstrated that SeedGerm is capable of scoring germination and measuring morphological changes automatically, for five major crop species and between different genotypes. The results show that SeedGerm could be employed to score germination frequency and seedling vigour, based on which the preformation of seed batches can be assessed. These traits were regularly checked by experienced seed engineers and scientists in order to provide certificates of seed germination and establishment performance in seed testing and seed insurance.

Hence, it is evident that SeedGerm has the potential to provide a replacement for manual assessment of germination frequency and radical emergence activities. Furthermore, as many traits measured by SeedGerm are highly correlated with seed performance and the effectiveness of post-harvest seed enhancement processes, SeedGerm is likely to contribute towards seed certification, guidance on sowing density, or even seed insurance in the future.

Besides routine seed testing of germination frequency, the applications of SeedGerm could also be expanded to seed vigour (i.e. how fast and uniform radical emergence) through the monitoring of morphological traits, which are important for estimating canopy closure, weed suppression, and crop yields, and for seed research.

Beyond existing traits analyses, continuous phenotypic analysis can extend our insights into the entire physiological procedure of germination, via increased understanding of phenotypic effects of individual seeds and seed batches under dissimilar treatments. Furthermore, we also set up a range of experiments to score germination rates and timing across a diverse panel of B. napus varieties to demonstrate the biological relevance of SeedGerm as a research tool to measure the effect of genetics. We showed that SeedGerm outputs can be used successfully for GWAS, identifying an association on B. napus chromosome A5 that explains the difference between high and low germinating varieties in the panel. Although the GWAS study identified associations over a 100kb region, the region does contain the gene BnaA05g27660D, a homologue of Arabidopsis AHG3, which is known to regulate ABA signalling during germination in Arabidopsis; this would therefore be a strong candidate for further study. The low germinating allele is only present in older spring varieties, including Bronowski and Duplo, suggesting that it has been consistently selected against by modern oilseed rape breeders. Hence, we believe that SeedGerm has great potential to have significant uses in seed germination scoring and seed testing, for both research and routine seed technology applications.

SMART IRRIGATION SYSTEM

Optimize irrigation water use

The primary agricultural irrigation system in Georgia is the center pivot irrigation system. Approximately 23,000 center pivot systems are located across the state, mainly below the Fall Line. Drip irrigation is also utilized in vegetable and orchard production statewide. Other forms of irrigation that are not as common include solid set sprinklers and traveling guns.

Some of the practices described below, such as remote control and irrigation scheduling, apply to any type of irrigation system. The other practices, such as retrofits and variable rate irrigation, are more specific to center pivot irrigation.

Enhance on-farm efficiency

Smart Irrigation tools range from simple and free to complex and costly. The best management practices described below are available to farmers in Georgia. Each practice includes a brief description, examples, an estimated cost, and approximate gains in water use efficiency or water savings. Note: water use efficiency refers to the amount of water pumped that is utilized by the crop (more crop per drop), and water savings refers to actual water conserved from reduction in groundwater withdrawal. The estimates below are not necessarily additive as each efficiency and savings scenario is dependent on individual field conditions and management decisions.

Arduino Based Smart Irrigation System using IoT

The hardware and software requirements of this project include Arduino UNO, soil moisture sensor, Wi-Fi module ESP8266, Arduino CC(IDE), Android studio, and MySQL, etc.

What is IoT?

The term "IoT" stands for the internet of things, can be defined as the interconnection between the individually identifiable embedded computing apparatus in the accessible internet infrastructure. The 'IoT' connects various devices and transportations with an help of internet as well as electronic sensors.

Arduino UNO Board

The Arduino UNO is one of the most used microcontrollers in the industry. It is very easy to handle, convenient, and use. The coding of this microcontroller is very simple. The program of this microcontroller is considered as unstable due to the flash memory technology. The applications of this microcontroller involve a wide range of applications like security, home appliances, remote sensors, and industrial automation. This microcontroller has the ability to be joined on the internet and perform as a server too.

Arduino is an open electronic platform for prototyping, based on flexible software and hardware that allows to make projects in a quick and easy way. Arduino can take information from the environment through their input mechanisms (pins), of a variety of sensors and can affect that surrounds controlling lights, motors, and other actuators.

The microcontroller on the Arduino board is programmed using the Arduino programming language and the Arduino development environment. Their projects can be run without connecting to a computer, while having the ability to do it and communicate with different types of software.

There are many different versions of Arduino boards. For the project presented in this work was selected the Arduino MEGA to control the sensors and actuators because it has a microcontroller ATmega1280. This board has 54 digital inputs / outputs, 16 analog inputs. Arduino Mega also facilitates communication with various aspects of the computer, another Arduino and other microcontroller if necessary.

In this project, the Arduino selected will be responsible for receiving all the information from sensors, and send it to a computer by serial communication. After receiving this information, it is processed by intelligent agents installed in a multi-agent architecture residing on a computer. Depending on the received data, and reasoning mechanism based on fuzzy logic, the Arduino turns on or off the irrigation system.

In the agriculture field, sensors are used like soil moisture. The information received from the sensors is sent to the Database folder through the Android device. In the control section, the system is activated using the application, this is finished using the ON/OFF buttons in the application. Also, this system is automatically activated when the soil moisture is low, the pump is switched ON based on the moisture content.

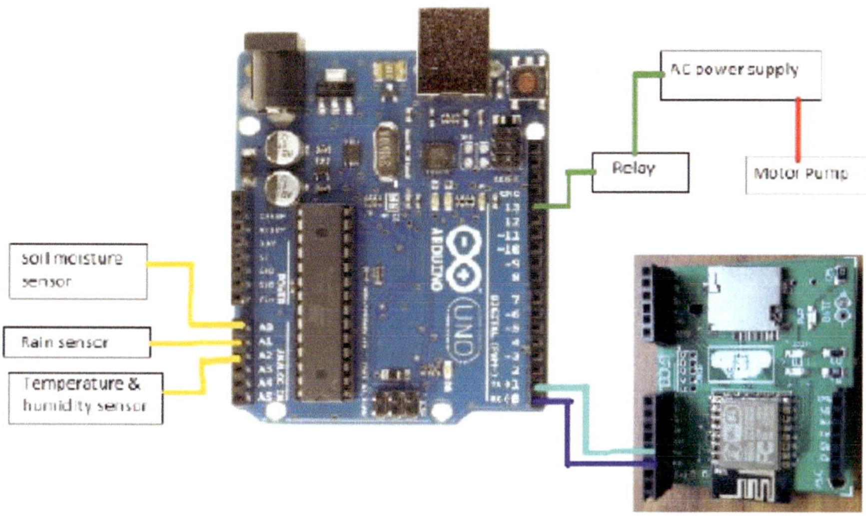

Fig. 7: Smart Irrigation System using IoT

The application has a feature like taking some time from the user and water the agriculture field when the time comes. In this system, there is a switch used to turn off the water supply if the system fails. Other parameters such as the moisture sensor demonstrate the threshold price and the level of water in the soil.

Further, this project can be enhanced by designing this system for large acres of soil. Also, this project can be incorporated to make sure the value of the soil and the expansion of harvest in each soil. The microcontroller and sensors are successfully interfaced and wireless communication is attained between a variety of nodes.

Also, further this proposed system can be enhanced by adding up machine learning algorithms, which are capable to study and recognize the necessities of the crop, this would aid the agriculture field to be an automatic system.

The inspections and outcomes tell us that this result can be executed for a lessening of water loss and decrease the manpower necessary for a field.

From the above information, finally, we can conclude that the hardware components of this system interfaces with all the sensors. The system is powered by a power source, and the system has been checked for watering an agriculture field.

Irrigation Retrofits

Evolution of efficiency

Farmers began implementing center pivot irrigation systems in Georgia in the 1970's, and the conventional models operated at high pressure with impact sprinklers spraying water from the top of the pivot mainline. Converting these systems from high to low pressure with spray-type sprinklers on drop hoses generates considerable water and energy savings.

By applying water at a lower pressure more directly to the soil surface, the low pressure system significantly reduces evaporation and wind drift losses.

Today's low pressure spray-type sprinklers are designed to produce a very uniform application of water droplets over the field. In combination with drop hoses, these sprinklers help to reduce losses to wind drift and evaporation by applying water closer to the crop in larger droplets.

High pressure impact sprinklers, in comparison, spray the water much higher in the air with smaller droplets. Also, low pressure spray-type sprinklers are unique in the fact that they operate between 10 to 30 pounds per square inch (psi) allowing them to have a lower energy requirement.

Software is helping farmers use water more efficiently

- The high pressure system has an average efficiency of 60%, due to water losses from evaporation, wind, and drift. Combined with the installation of end-gun controls to keep irrigation inside the field boundary and the repairing of leaks in the system, retrofitting a pivot may increase water use efficiency to 80-90%.
- Putting precision into irrigation. Farmers are turning to Trimble's Irrigate-IQ water management solution to enable GPS-guided irrigation systems that precisely control every nozzle for highly targeted water application. When implemented, they can reduce water consumption by 30 percent.
- Using sensor data to precisely deliver the right amount of water. In California, a state continuously hit by drought, agriculture consumes about 80 percent of the state's water. California's thirstiest crops are almonds.
- Growing a single almond consumes around a gallon of water. But now, by connecting wireless sensors across a farm, data can be collected, crunched in the cloud, and used to control an irrigation system that drips in water at just the right amount - using 20 percent less water than traditional irrigation techniques.

Fig. 8: A center pivot irrigation system with high pressure sprinklers

Fig. 9: A center pivot irrigation sprinkler with low pressure drop-nozzle sprinklers

- Saving an aquifer by using water with digital precision. In Kansas, the aquifer is dropping by as much as two feet per year in some counties.35 Studies suggest that if existing trends continue, 70 percent of their water will be depleted by 2060.
- As a result, the Kansas Water Office partnered with forward-thinking farmers who are taking advantage of new technologies to identify ways to more efficiently use scarce water resources. T&O Farms combined soil sensors, precision weather predictions, and a drip irrigation system to cuts its water use by 33 percent without sacrificing yield. Across the whole season, the savings translated into about three inches of water saved for the entire crop — water savings that can be readily replicated throughout Kansas.
- Delivering water just to the vine, to improve the quality of California's wine. California is also home to a $35 billion wine industry, which is rooted in its ability to consistently produce quality grapes. To maximize the quality of its grapes, E. & J. Gallo Winery turned to IBM's Watson to develop an intelligent irrigation system that uses weather reports and remote sensor data to deliver precise amounts of water to each vine to increase the quality of its grapes. Because of this tailored watering, the winery reduced its water use by 25 percent, while also improving the quality of its wine.
- Improving the quality and quantity of crops in the developing world. These technologies can be especially powerful when used abroad. For example, the millions of small-scale farmers in Kenya traditionally don't irrigate for their crops. IBM has stepped up to help Kenyan farmers optimize crop growth and improve food security by managing their water more effectively using Internet of Things (IoT) technologies.
- The IBM project EZ Farm uses sensors to help farmers understand how much water is in the soil and measure the amount of water in local water tanks. This data is then combined with weather data to plan irrigation times to optimize crop growth, extend the crop cycle, and improve the quality and quantity of crops.

COMPUTER-AIDED FERTILIZATION SOFTWARE

Precision agriculture presupposes accurate dosage of mineral fertilisers in quantities adjusted to the nutritional needs of the plants. Modern spreaders can dose fertilisers changing the dose in different parts of the field taking advantage of GIS maps and GPS navigation. This solutions allow to utilise fertilisers in an effective manner; however, due to high prices of the machines and devices, they are, as a rule, not used in small farms.

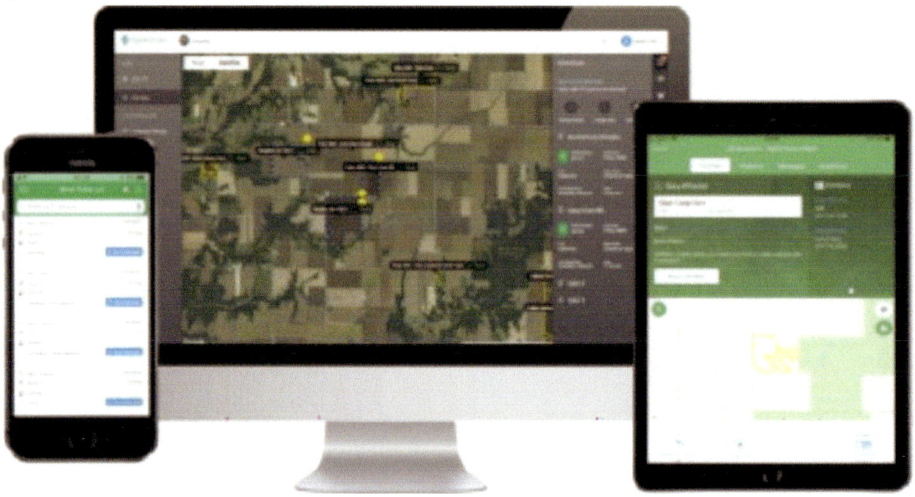

Fig. 10: Agvance Blending was updated to integrate with other Agvance agronomy tools.

Farmers who only have several hectares of land calculate the dose of fertilisers in a simplified way, which may imply that plants will not receive the desired amount of nutrients or that too much fertiliser will be applied. The correct fertiliser dose selection should be based on investigations into the soil content of phosphorus, potassium and magnesium, as well as the soil pH and also the need for liming. These as well as other specialist analyses can be performed at the Regional ChemicalAgricultural Stations or by companies offering such services. The analyses can be supplemented with fertilising recommendations.

Functions enabling calculation of the correct dose of mineral fertilisers are available in specialist software dedicated exclusively to this purpose and in some selected programs intended for full farm management. Such programs include NawSald, Naw-3 and InfoPlant offered by the Institute of Soil Science and Plant Cultivation and also by other suppliers.

Blending software is critical. It not only improves blending, inventory, and billing procedures it dramatically improves grower satisfaction. Wholesalers and retailers can offer growers more accurate blends, faster. Retailers and wholesalers who use ICS share how they've been able to grow thanks to the better experience they're able to offer growers in terms of speed, responsiveness, and billing and blend accuracy.

Smart fertilizers

Slow-/controlled-release fertilizers

The slow- or controlled-release fertilizers are those containing a plant nutrient in a form, which either (a) delays its availability for plant uptake and use after application, or (b) is available to the plant significantly longer than a reference "rapidly available nutrient fertilizer" such as ammonium NO - or urea, ammonium phosphate or potassium chloride. There is no official differentiation between slow release and controlled-release fertilizers. However, the microbially decomposed N products, such as urea-formaldehydes, are commonly referred to as slow-release fertilizers, and coated or encapsulated products as controlled-release fertilizers.

Delayed availability of nutrients or consistent supply for extended time periods can be achieved through a number of mechanisms. These include semipermeable coatings for controlled solubility of the fertilizer in water, protein materials, occlusion, chemicals, slow hydrolysis of water-soluble compounds of lower molecular weights, and some other unknown means. Other options include utilization of semipermeable materials and sensors of chemical or biological origin within the fertilizer.

These are advanced materials, whose physical or chemical properties can change in response to an external stimulus such as temperature, pH, and electric or magnetic fields. Many coating materials can be used to slow nutrient release, including natural materials such as clays and nanoclays (e.g., allophane), and nondegradable (polysulfone) and biodegradable polymers (e.g., alginate beads). Several studies have shown that urea-based coatings can have variable efficiencies depending on the material used. In addition, they can be highly expensive in some cases, pollutants, or toxic (i.e., polymer-coated urea) and mostly difficult to degrade with potential environmental impacts (accumulation) as the major concern.

However, these polymers may need some modifications to be included into the coating design because of their hydrophilic properties and their weak coating barrier, for example, in the case of starch.

As a consequence, formulation with environmentally safe and biodegradable coating materials seems to be necessary. In this sense, natural polymeric carbohydrates appear as an alternative to nonbiodegradable materials acting as permeable or impermeable membranes with tiny pores in slow-release fertilizers (e.g., urea). These highly degradable materials have also received attention because of their low cost and low environmental damage due to biodegradability and low accumulation in the environment.

Biofertilizers

One group of microorganisms beneficial for plant growth is PGPR, a heterogeneous group of bacteria that can be found in the rhizosphere, at root surfaces, and in association with roots. These bacteria have several functions, including production and regulation of phytohormones, release of nutrients to plants (e.g., P-, N-fixation, siderophores, among others), and control of phytopathogens (production of antibiotics and siderophores).

Phosphobacteria, phytate-mineralizing bacteria, and phosphatesolubilizing bacteria have been commonly isolated from soil and proposed as inoculants for agricultural improvement. For example, a large diversity of microbes in Andisols under pastures and cereal crops are capable of mineralizing phytate.

They may be used to develop bacterial or enzyme systems as biofertilizers to overcome the limitations of conventional fertilizers in acidic soils, as well as for developing added value products from agricultural wastes. To this end, researchers studied the effect of enzyme–nanoclay complexes on P availability of composted cattle dung and showed that it increased the inorganic P content. Moreover, they suggest that inoculation of cattle manure with enzyme–nanoclay complexes enhances the organic P cycling and P nutrition of plants grown in P-deficient soils. Low N acquisition by plants is a limiting factor in agricultural ecosystems, and there is interest in using N2-fixing bacteria as an alternative to conventional fertilization. Diazotrophic bacteria are capable of converting atmospheric dinitrogen (N2) into NH3, which can be used by plants.

Among them, a number of free-living soil bacteria are considered to be PGPR because of their competitive advantage in C-rich and N-poor environments. Free-living N2-fixing bacteria have been considered as an alternative to conventional N fertilizer for promoting plant growth, and several research studies reported significant increases in grain and shoot biomass yield from plants inoculated with freeliving diazotrophic bacteria.

Moreover, researchers reported that under intensive wheat rotation at Avon, South Australia, free-living N2-fixing bacteria contributed 20 kg N per ha per year, which represented 30%–50% of total crop requirements. This response was attributed to a combination of factors including enhancement of root development, production of growth regulators, and N2-fixation. However, it is well known that bacteria directly inoculated in the soil system could be adversely affected by competition with native microorganisms, unfavorable physicochemical conditions, and fluctuating pH and temperature.

Encapsulating microorganisms in carrier materials (bioformulation) is designed to protect them during storage and from adverse environmental condition (pH, temperature, etc.), thus ensuring a gradual and prolonged release. A wide range of microorganisms have been investigated, and a framework for selecting suitable organisms for specific purposes has been developed.

Materials suitable for immobilization and preservation of bacteria include alginate gels, synthetic gels (Sol–Gel), polyacrylamide, agar and agarose, polyurethane, vermiculite, and polysaccharides. In addition, composite materials based on biodegradable polymer clay or nanoclays are being studied, including nanocomposites.

For example, encapsulation of free-living diazotrophic bacteria has been considered as one of the possible alternatives for inorganic N fertilizer for promoting plant growth and crop yield.

Fighting Pests and Protecting Crops While Reducing Harmful Chemical Use

Each year, as much as 40 percent of the world's potential harvests are lost to damaging insects, weeds, and plant diseases.38 At the same time, chemical runoff from the overapplication of pesticides, herbicides, and fertilizers can affect the health of our oceans, rivers, and streams. In fact, farmers who don't yet use precision fertilizer technology are estimated to be overfertilizing 40 percent of their fields, and still seeing yield loss on 10 percent of their fields. That's in part because an estimated 95 percent of chemicals sprayed miss their intended target, wasting the expensive agricultural chemicals and raising the toxicity risk for both people and wildlife. To solve the problem, farmers are turning to software-enabled precision technologies to reduce waste by delivering just the right amount of chemical precisely where it's needed.

- **Applying fertilizer in calibrated doses.** Software-enabled technologies now help apply fertilizers in more calibrated doses to meet specific soil and plant nutrient needs, dramatically cutting fertilizer and other chemical use. For example, Trimble's Field-IQ software is a fertilization control system that helps control variable rate sprayers aimed at precisely managing a grower's inputs. Using an integrated system of hardware (precision nozzles, boom adjusters, and spinner speed controls) and its innovative software, the system provides a more accurate seeding and fertilization system to the grower.
- **Using soil sensors to improve nutrient uptake.** Software is also helping improve crop health by optimizing soil pH levels, which is essential for nutrient uptake that can reduce the amount of fertilizer needed and directly impact yields. The growers at Costa Farms in Florida partnered with Microsoft to create a better way to measure and regulate pH throughout the day and across the lifecycle of its plants. Based on the Azure IoT system with pH sensor devices, Costa Farms is now able to not only automatically measure pH levels, but dramatically increase its yield using fewer chemicals.
- **Advancing weed-whackers to spot and zap weeds more efficiently.** New technologies also create opportunities to kill weeds more accurately. Robotic weed-whackers with image recognition software can recognize weeds and zap them with a laser or apply a micro-dose of herbicide directly to a weed's leaf. These devices can cut herbicide use by as much as 99.99 percent.
- **Crowd-sourcing caustic critter solutions.** Software also enables new ways to tackle infestation problems. When some local farmers had lost up to 90 percent of their crops to fruit flies, Greenwood Orchards turned to Microsoft's

cloud technology to fight the critters. Using traps tied to GPS coordinates, they developed a crowdsourced application that created a community-wide infestation monitoring system to detect the early warning signs and enable early action to prevent devastating consequences.

8

AEROPONIC SYSTEM

INTRODUCTION

As technology develops, it will bring great advantages to human beings in all areas of life. Especially, the development of network technology is changing many areas of our lives such as smart homes, smart cities, smart grids, autonomous cars, and the industrial internet. Smart agriculture systems are an area that would benefit from this technology. By adapting the recent innovations in technology for farming it would add many benefits for farmers and reduce our carbon footprint.

The Internet of Things (IoT) is an emerging technology that provides internet connectivity and powerful data analytic capabilities to create a system in which objects in the physical world could be connected to the internet by sensors. However, smart farming is still in its infancy in regard to using IoT devices for agriculture. Aeroponics farming is an efficient and effective process for growing plants without using soil as a medium as well as with little water. In this paper, we propose new automated aeroponics system using IoT devices that will help the farmer by increasing productivity in farming. To the best of our knowledge, this is first work to design and implement of automated aeroponics system using IoT devices.

The aeroponic system is one of the techniques of the soilless culture, where the plant grows in the air with the assistance of artificial support instead of soil or substrate culture. It is an air-water plant growing technique where lower portions such as the roots of the plant are hanged inside the growth chamber under complete darkness in controlled conditions. However, the upper portions of the plant such as leaves, fruits, and crown portion are extending outside the growth chamber. Usually, the artificial supporting structure (plastic or thermofoam) is provided to support and divide the plant into two parts (roots and leaves). In the system, plant roots are openly exposed in the air and directly irrigated with a small droplet size of the water nutrient at interval basis. The nutrient solution is supplied through different atomization nozzles with or without high air pressure. Moreover, several studies considered aeroponics as a modern-day agricultural activity which is practiced in an enclosed growth chamber under entire controlled conditions, as it could eliminate the external environmental factors as compared with traditional agriculture activity. Hence, it is no longer dependent on large-scale land use, and it could be set up in any place, a building that has lifted global climate without considering the current climate such as rainy season and winter. Researchers reported that atomization nozzle uses the tiny amount of the water nutrient solution and provides an excellent growth environment for the plant.

Researchers said it is a modern-day agricultural research tool which provides several agricultural research opportunities for a researcher with significant results by providing artificial growth conditions. However, Table 1 shows the essential monitoring and control parameters in the aeroponic system. Researchers studies discovered that aeroponics contributes to the advances and developments in many areas of plant root studies. It provides an excellent chance for plant researchers to deeply study the behavior of plant root under different conditions and without any complications. Until now, many researchers had conducted plant root research and experimental studies root response to drought, effects of different oxygen concentrations on plant root development, root microorganism, arbuscular mycorrhizal fungi production, and legume-rhizobia interaction. Furthermore, studies also practiced the technique by growing vegetables, fruits, herbs, and medicinal root-based plant such as tomato, potato, soybean, maize, lettuce, Anthurium andreanum, and Acacia mangium.

DESIGN AND IMPLEMENTATION OF AEROPONICS SYSTEM

In this section, we introduce our aeroponics system using IoT devices including its three main components, mobile interface, service platform, and IoT devices with each sensor.

Mobile application

We designed a mobile interface that a user can easily check every sensor's value in each aeroponics system through a service platform. This service platform stores the sensor's data such as temperature, humidity, pH balance, nutrient levels, LED lights and water level. Referencing Fig. 1, the user can view the current sensors' data on the service platform of the aeroponics system. For the service platform, the current web technologies HTML5, CSS Flexbox, Javascript, and SVG are used.

SERVICE PLATFORM

We designed and implemented the service platform to provide an interface to view the sensor data of the aeroponics system. In Fig. 2, the service platform provides a web interface to access data using a mobile device or client, and provides a database service to save the gathered information from each sensor using IoT devices. For the web service, we installed Apache 2.4.26, MariaDB 10.1.25, and PHP 7.1.7 on Linux system.

Table 1 Basic monitoring and control parameters in the aeroponic system

No. Parameters	Common value	Instruments
1Nutrient atomization	Mist/spray/aerosol/droplet size at high pressure from 10 to 100, low pressure from 5 to 50, and ultrasonic foggers from 5 to 25 microns, respectively	Atomization nozzle (high and low low pressure, atomization foggers)
2 Growing medium	Plant holder	Any artificial root supporting structure
3 Desirable pH of the nutrient solution	The pH value depends on the cultivar (onion 6.0–7.0, cucumber 5.8–6.0, carrot 5.8–6.4, spinach 5.5–6.6, lettuce 5.5–6.5, tomato 5.5–6.5, and potato 5.0–6.0)	pH measuring device
4 Desirable EC of the nutrient solution	The EC value depends on the cultivar (onion 1.4–1.8, cucumber 1.7–2.2, carrot 1.6–2.0, spinach 1.8–2.3, lettuce 0.8–1.2, tomato 2.0–5.0, and potato 2.0–2.5 ds·m"1)	EC measuring device
5 Humidity	Provide 100% available moisture	Humidity measuring device
6 Temperature	Optimum 15°C–25°C and should not increase to 30°C and less than 4°C	Temperature measuring device
7 The light inside the box	The light inside the growth box must be dark enough	Cover the growth chamber with locally available material
8 Atomization time	Depends on the cultivar growth stage	Manually operating the system with timer
9 Atomization interval time	Depends on the cultivar growth stage	Manually operating the system with timer

Fig. 1: Graphic user interface for mobile application

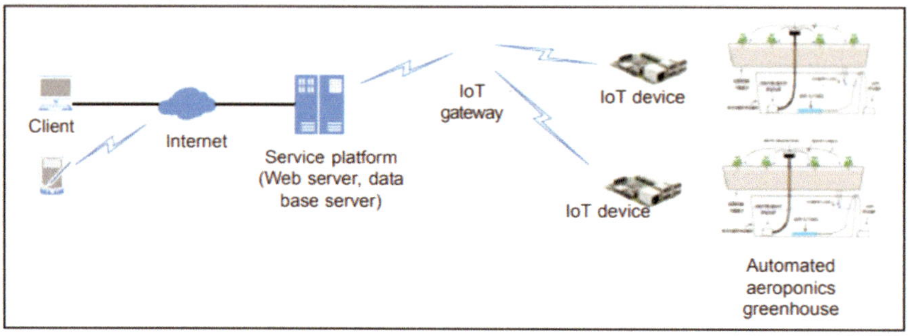

Fig 2: Automated aeroponics system using IoT devices

IOT DEVICE CONNECTED ON SENSORS

We implemented a module to gather each sensors data using IoT devices. Raspberry PI Zero is one of the IoT devices. In this research, we use Raspberry PI Zero to gather information (pH balance, temperature, and humidity) from the sensors to control a water pump and dosing pump to add nutrients and water. The gathered information is saved into a database server by sending a SQL query. GPIO pins on the Raspberry PI Zero connects the DHT11 temperature and humidity sensor, Atlas scientific pH probe and EZO Circuit, water level sensor, LED lights, submersible pump and dosing pumps for the delivery of nutrients. Figure 3 shows the flow chart of the IoT device system

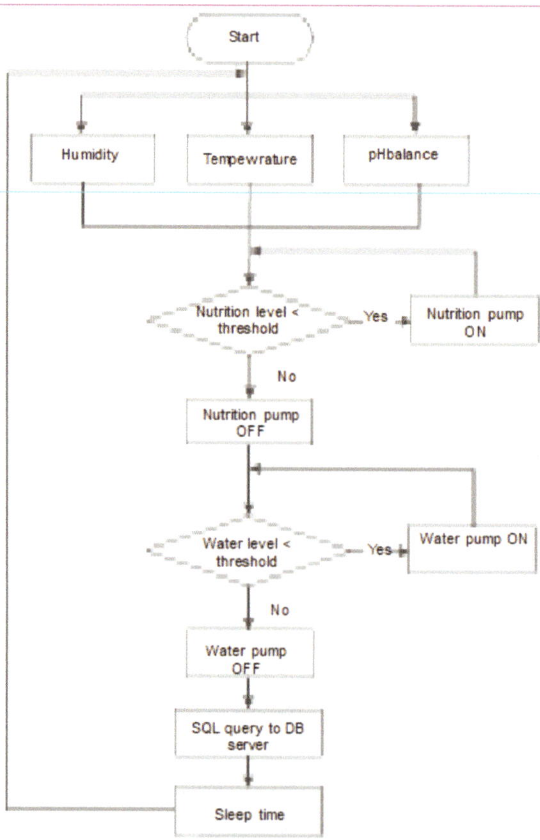

Fig. 3: Flow diagram of IoT device

Each IoT device gathers sensor data (temperature, humidity, and pH balance) during a certain time. After checking nutrition levels if the nutrition level is below a certain threshold value the Raspberry PI Zero will trigger a relay. The connected dosing pumps start to work to add nutrition in the aeroponics system. If the nutrition level is over a certain threshold value it will stop adding the nutrients in the aeroponics system. The submersible pump process works the same as the dosing pump process.

PRESENT STATUS OF THE AEROPONIC SYSTEM

The aeroponic system is one of a holistic production management method in agriculture which promotes and improves agroecosystem, health, and biodiversity. The system has a paramount reputation in the horticulture department, because of its implications on the economic and technical aspects in the agriculture. Among all agriculture systems, only the aeroponic system is receiving the full attention of farmers, policymakers, entrepreneurs, and agricultural researchers. The grower could reduce the requirements of chemical inputs including fertilizers, herbicides, pesticides, and other agrochemicals. The grower could obtain higher cultivated plant yield and quality as compared with other growing methods.

However, the aeroponic system is labour-intensive. It offers many opportunities for the farmers to increase rural employment. The farmer can grow a plant in their homes by providing artificial growth environment. Researchers reported that nowadays many families are practicing the aeroponic system on their terrace. Besides, several countries of the world are using aeroponics for making an expansion in nourishment creation managing the monetary issues and making the nation naturally amicable to have their particular food supply. While some years ago, the use of the aeroponic system was limited almost around the world. At present, the system is acquiring more attention from the farmers and several countries are being effectively adopting as an economical and environmentally friendly vegetable and fruit growing system.

However, it is practiced in following countries: Abu Dhabi, Australia, Bhutan, Bolivia, Brazil, Bangladesh, Burkina Faso, China, Canada, Colombia, Ecuador, Egypt, Ethiopia, France, Germany, Ghana, Greece, Indonesia, Italy, India, Iran, Japan, Israel, Kenya, Korea, Malaysia, Mongolia, Malawi, New Zealand, Nigeria, Peru, Philippines, Poland, Russia, Rwanda, Saudi Arabia, South Africa, Spain, Singapore, South Korea, Slovakia, Sri Lanka, Taiwan, Thailand, Uzbekistan, and Vietnam. Besides, attempts are made to represent the system in other countries of the world.

KEY PROBLEMS AND DIFFICULTIES OF THE AEROPONIC SYSTEM

Aeroponic cultivation is performed in an outdoor and indoor installation and or in a greenhouse under controlled conditions. It may be carried out within a facility that includes the provision of light for plant growth, the centralised delivery of nutrient solution, and electrical power. The growing plants are set in a growth chamber and periodically soaked with nutrient solution small mist ejecting through atomization nozzle. In addition, the aeroponic system gives the chance to control the entire growth chamber environment precisely. The aeroponic system is the modern technique of the agriculture which is still under development. Until now, limited studies have been performed, and conducted studies concluded that the system has some problems and issues. Studies suggested that aeroponics is performed without soil or any solid media; thus, the main observed problems are water and nutrient buffer, any failure of the water pumps, nutrient solution distribution and preparation, atomization nozzle clogging, and so on, which lead to rapid death of the grown plant. Researchers reported that the aeroponic system provides better control of the plant growth and nutritional availability and prevents the plant from various diseases and root rot. However, during plant growth from sowing to harvest time, the methods adopted in the aeroponic system require a little hand-operated contribution, interference regarding physical presence, and expertise in domain knowledge of plants, environment control, and operations to maintain and control the growth of the plant.

Moreover, there is a requirement to sustain and keep retain the nutrient solution parameters which include nutrient temperature, pH, and EC concentration in a narrow range of preferred values for optimal growth. If these parameters drift outside the desired range, it will create several problems for plant growth. In addition, some supplemental parameters can adjust to optimise the plant growth further. The additional parameters are atomization time, atomization interval time, air temperature, relative humidity, light intensity, and carbon dioxide (CO2) concentration which make the system complicated and time-consuming with high human energy and with the higher level of expert training and skill for operating the system.

However, the grower has the responsibility to control and monitor the fluctuations of the above parameters in the desired range to achieve the suitable growth conditions for the specific plants. A failure to accurately control and monitor the parameters could significantly affect the growth of the plant and cause financial loss. If any component failure occurs while the operator is not present on site, it may be detected too late to prevent harm, because systems generally include some automated means for periodically providing nutrient mist to the plant roots, refilling a nutrient reservoir, and managing light cycles and intensity.

Therefore, the aeroponic cultivation considered hitherto to be somewhat unsuitable for the local grower and due to the above reasons and it is not common to find an installation. However, the main reason for the low acceptability of the aeroponic system is not a cost, but the main drawback is the amount of attention required of the grower with a high level of expertise and judgment. For the above-discussed reasons, more sophisticated and advanced monitoring techniques have implemented in the aeroponic system for early fault detection, real-time monitoring, and control and automation of the system. Hence, it would be advantageous to use artificial intelligent tools in the aeroponic system to detect fault and diagnosis problems on time. Thus, it could help to avoid rapid damages to grown plants and help to fully automate the aeroponic system.

THE AEROPONIC SYSTEM AND SENSOR NETWORK

In recent years, early fault detection and diagnosis using an intelligent agricultural monitoring system is considered as the best tool to monitor plant without any complicated operations and laboratory analysis which required domain expertise and extensive time. The development of these convenient features has attracted much attention in the agriculture. The system is based on a wireless sensor network which comprises of a data server, a wireless convergence node, a plurality of wireless routers, and a plurality of wireless sensor nodes. However, the wireless sensor nodes are used as the signal input of the intelligent agricultural monitoring system and are used to collect each selected parameter of farming operations to be monitored.

Researchers stated that wireless sensor network-based systems could be a significant method to fully automate the agriculture system, because the sensors provide real-time significant information and believed to eliminate the considerable costs of just wiring. Another study stated that in agriculture, sensor network technique helps to improve existing systems installed in the greenhouse efficiently and smoothly by forwarding real-time collected information to the operator through the radio signals. The system optimises the transmission protocols more accurate and quick and maximises the application of energy to save the energy and reduce the consumption. Researchers suggested that the utilisation of artificial intelligence techniques in the aeroponic systems could lead not only to find early fault detection but also to fully automate the system without any or small interventions of human operators. The aeroponic system could gain more popularity among local farmers by deploying this technique in a system for monitoring and controlling purpose. However, it will conserve resources and minimise impacts on the environment. The farmers could start to understand their crops at a micro scale and able to communicate with plant through accessible technology.

Therefore, in this section, we explored how wireless sensing technologies wove into the aeroponic system. Thus, the primary motivation of this review article was to provide an idea about different intelligent agriculture monitoring tools used for early fault detection and diagnosis for plant cultivation in the aeroponic system. Additionally, it would be helpful for the local farmer and grower to provide timely information about rising problems and influencing factors for successful plant growth in the aeroponic system. The adoption of the intelligent agriculture monitoring tools could reduce the concept of unsuitable for the amateur.

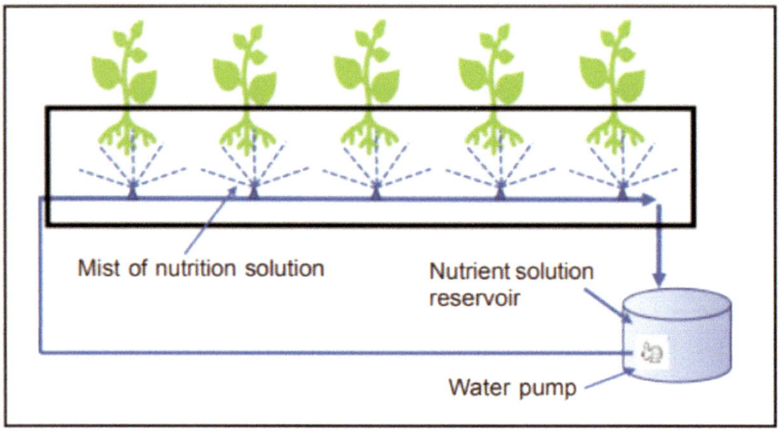

Fig. 4: Aeroponic cultivation control system.

Number of Sensor Nodes and Input Parameters

At present, the utilisation of different sensor techniques is almost possible in every field of life due to the sharp progressions in the currently available technologies. Moreover, the sensor is a device that has capabilities to measure physical attributes and convert them into signals for the observer. A WSN (wireless sensor network) traditionally consists of a few to dozens and in some cases thousands of the sensor nodes which are connected to one or more sensors. Generally, it includes a BS (base station), which acts as a gateway between the WSN and the end users. Each sensor node is consisting of five main components, which are a microcontroller unit, a transceiver unit, a memory unit, a power unit, and a sensor unit. Each one of these components is a determinant in designing a WSN for deployment. Furthermore, the microcontroller unit is in charge of the different tasks, data processing, and the control of the other components in the node. Through the transceiver unit, a sensor mode performs its communication with other nodes and other parts of the WSN. It is the most dominant communication unit. The memory unit is another important part of the WSN system, which is used to store the observed data. The memory unit could be RAM, ROM, and their other memory types flash or even external storage devices such as USB. Lastly, the last one unit is the power unit. It is one of the critical components of the system which is for node energy supply.

However, the power unit could be any source; it can store in batteries (most common) rechargeable or not on in capacitors. In addition, for extra power supply and recharging the power unit, the available natural resource could be used. The natural sources induce solar power energy in forms of photovoltaic panels and cells, wind power energy with turbines, kinetic energy from water, and so on. Last but not the least is the sensor unit, which includes several sensors for parameter measurements such as temperature, humidity, carbon dioxide, methane, and carbon monoxide. However, in the aeroponic system, the total required number of sensors and actuators depends on the size and requirement of the operator.

Sensor Types and Monitoring Parameters

In this review study, we reviewed the previous work done on the aeroponic system using wireless sensor network technique. We found that the primary objective of a wireless sensor network system for the aeroponic system is to control the growth chamber climatic condition as per the crop data sheet. However, the basic principle of the aeroponic system is to grow the plant by suspending in the closed, semiclosed, or dark environment in the air with artificial provided support. In the system, the plant stems, leaves, and any fruit grow in a vegetative zone above the suspension medium, and roots dangle below the suspension medium in an area commonly referred as a root zone.

Generally, closed cell foam is compressed around the lower stem and inserted into an opening in the aeroponic growth chamber, which decreases labour and expense. However, the trellising is used to suspend the weight of cultivated plant. Ideally, the environment is kept free from pests and diseases so that the plants grow healthier and more quickly than other plants grown on techniques.

Furthermore, the key to the success and high yields of the air gardening is a scientific grade monitoring of the conditions and accurate control of the growing environment. Each plant yields and needs a different environmental condition for growth. However, the plant growth is mainly influenced by the surrounding environmental and climatic variables and the amount of water and the fertilizers supplied by irrigation.

There is a requirement to monitor and control liquid nutrient parameters in a narrow range of preferred values for optimal growth. The parameters include nutrient temperature, pH, and EC concentration. If the parameters drift outside the desired range, the plants can harm. Besides, there are some additional parameters which can be adjusted to further optimise growth, such as air temperature, relative humidity, light intensity, and carbon dioxide (CO2) concentration. Researchers reported that the one solution to solve the problems of monitoring and controlling the growing conditions in the space environment is by applying some sensors. The sensor can detect and monitor a number of parameters such as temperature, humidity, light intensity, O2 and CO2 levels, direction, and wind speed. Aside from the sensors, there is also a requirement for the actuators to distribute nutrients and waters to plant roots or lower stems (Fig. 5). The sensor collects the information of the various environmental conditions and forwards the signals to the actuator to take place and produce the outcome for the collected information to know the status of that parameter. The actuator can control the environment changes. The sensors store information that analyzes the environment and identifies the location, object, people, and their situations. The sensor provides multiple contributions in various domains that depend on a variety of attribute and variant in time

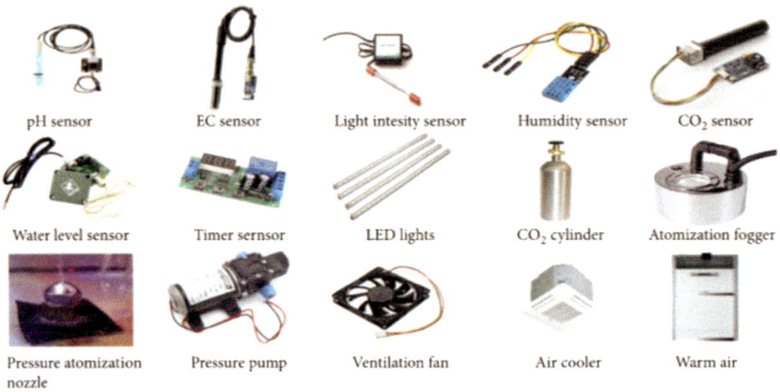

Fig. 5: Sensors and actuators used in an aeroponic system.

Temperature Sensor

In the aeroponic system, the temperature is one of the critical factors significantly determining plant growth and development. A reduction in temperature below the optimal conditions often results in suboptimal plant growth. A different cultivar requires a different temperature level for the photosynthesis process and growth, which can advance the plant growth stage. It will eventually bring us substantial economic benefits. In the aeroponic system, the optimum growth chamber temperature should not be less and more than 4 and 30°C, respectively, for successful plant growth. The temperature fluctuations of aeroponic growth chamber can significantly affect the root growth, respiration, transpiration, flowering, and dormant period. Therefore, the temperature sensors can be used to monitor the temperature fluctuations of the aeroponic system. At present, temperature sensors are used in many applications like environmental controls, food processing units, medical devices, and chemical handling. The temperature sensor is a device mainly composed of thermocouple or resistance temperature detector. The temperature sensor measures the real-time temperature reading through an electrical signal. The sensors collect the data about temperature from a particular source and convert the data into an understandable form for a device or an observer. The temperature sensor accurately measures temperatures slower changing from critical applications such as facilities or rooms and sends them to the user's webpage.

Humidity Sensor

Aeroponics is the technique of cultivating plant by providing the water nutrient small spray in the air. Thus, the humidity is another important parameter of aeroponic growth chamber environments, and its control is recognised to be very important for significant plant growth. In the aeroponic system, the plant gets all available moisture in the growth chamber. Moreover, if the growth chamber has too high or less moisture content, both conditions will create many problems for the plant. Accordingly, an accurate and precise means of testing moisture content in the growth chamber will help farmers to monitor their crops and provide a suitable growth environment for the plant. Researchers reported that a humidity sensor is a device that detects and measures water vapour present in the air within a room or enclosure. At present, humidity sensors are widely used in medicine, agriculture, and environmental monitoring. However, the most commonly used units for humidity measurement are relative humidity. The development of humidity sensors has shown remarkable progress because of using various types of sensing materials in recent years. The sensing materials used in humidity sensors can classify into ceramics, polymers, and composites. The humidity sensor could be placed in the growth chamber to maintain the moisture level. If the moisture level becomes less than the plant requirement, the sensors will forward the signals to atomization nozzles to perform their work.

Light Intensity Sensor

As we know, all vegetable plants and flowers require large amounts of sunlight, and each plant group reacts differently and has the different physiology to deal with light intensity. Some plant performs well in low light intensity and some in high light intensity. However, the aeroponic system implements in indoor conditions, so it is necessary for the farmer to provide sufficient light quantity of at least 8 to 10 hours for a day to grow the healthy plant. The artificial lighting is a better option to present enough intensity to produce a healthy plant. In the conventional aeroponic system, the control of the light quantity present in the growth chamber is mostly done by farmer through observing the plant condition. However, it is a time-consuming and challenging task for the farmer to provide the required light concentration accurately. It could be a better option to use intelligent agriculture techniques to monitor the light intensity in the aeroponic system. The intelligent agriculture techniques mean using the sensor system to control the light intensity. The light sensor is an electronic device which is used to detect the presence or nonpresence of light and darkness. There are several types of light sensors including photoresistors, photodiodes, and phototransistors. These light sensors distinguish the substance of light in a growth chamber and increase or decrease the brightness of light to a more comfortable level. Light sensors can be used to automatically control the lights such as on/off. By adopting the sensor network in aeroponics, the farmer could be able to monitor light intensity without any human interference. Because the sensors will perform all work such as if the light intensity in the growth chamber will be less than the required light quantity for plant growth, the sensor will automatically forward the signal to the LED light to turn on until the light quantity reaches to the desired level.

CO_2 Sensor

The appropriate oxygen concentration in the root environment is crucial to keep the root metabolism in nutrition solution. The available oxygen concentration for the root environment is a hugely significant factor since low concentrations affect the root respiration, nutrient absorption, and, consequently, the plant growt. Thus, the CO_2 sensor could be used to monitor the carbon dioxide fluctuations in the aeroponic growth chamber. A carbon dioxide sensor is an instrument which is used for the measurement of carbon dioxide gas concentration. Researchers reported that CO_2 sensors are used to measure indoor air quality in a building to perform demand-based ventilation. However, the CO_2 sensor data measuring range is in between 500 and 5000 parts per million. There are two main types of the CO_2 sensors which include nondispersive infrared carbon dioxide sensors (NICDS) and chemical carbon dioxide sensors (CCDS), whereas the NICDS detected CO_2 in a gaseous environment by its characteristic absorption and composed of an infrared detector, an interference filter, a light tube, and an infrared source. However, the CCDS of sensitive layers are based on polymer or heteropolysiloxane with low-energy consumption.

Water Level Sensor

The aeroponics is the method of the plant cultivation by providing a small mist of the nutrient solution in the growth chamber. Thus, there is no any use of soil; just water is required to cultivate the plant throughout the germination to harvest time. Therefore, the water nutrient solution reservoir is one of the major components of the aeroponic system which should be monitored throughout the growth period. In the conventional aeroponic system, the farmer checks the water nutrient level in the nutrient solution reservoir, and if he founds water level less than the desired level, he maintained accordingly. However, by adopting the precision agriculture techniques, the farmer will be able to monitor and control water nutrient level through the intelligent methods such as wireless sensors. The water nutrient level sensors detect the liquid level in the reservoirs and facilitate operator in collecting water nutrient level data in real time. The sensors will alert the operator about any potential property damage that results from any leaks and also allowing to know when a container is nearing empty.

EC and pH Sensor

In the aeroponic system, the plant productivity is closely related to nutrient uptake and the EC and pH regulation of the nutrient solution. The EC and pH concentration of the nutrient solution affects the availability of the nutrients to plants. The pH and EC concentrations are controlled to prevent barrier growth. Their measurement is essential because the solubility of minerals in acidic, alkaline, and ion concentration of all the species in solutions is different and the solution concentration changes with solubility.

The unmonitored EC and pH concentration of the nutrient solution will quickly lead to a situation where plants cannot absorb the essential nutrients, if not corrected this will eventually lead to harmful plant growth and poor productivity. Thus, the EC and pH concentration of the nutrient solution is a critical parameter to be measured and controlled throughout the plant growth. Moreover, in the conventional aeroponic system, the EC and pH value of the nutrient solution is mostly monitored manually by performing laboratory analysis or using advanced equipment which is a time-consuming process.

For instance, when the EC of the nutrient solution decreased or increased, the control of nutrient solution concentration is mostly achieved by adding more high concentration nutrient solution or the fresh water, respectively, to the nutrient solution to maintain the EC level to the prescribed target range. Similarly, for pH, an acid solution and an alkali solution are used to control the pH fluctuation of the nutrient solution within a specified target range.

However, these conventional methods are time-consuming and challenging task for the farmer to maintain the EC and pH value at the desired range accurately. In addition, the EC and pH sensor could be used to deal with the above challenges.

SENSOR WORKING PROTOCOL IN THE AEROPONIC SYSTEM

Today, the world demands automatic tools to do most of the work for them without bothering its user for doing some task. So, the concept is all about a very high level of automation system which will be independent of its users to a very great extent, reduce human efforts, and save all kinds of resource utilisation, as monitoring and controlling will be done by computers leaving very few easily manageable tasks for humans, and it will interest more people to join this field.

Moreover, the monitoring and control system for the aeroponic system mainly consists of following sections which include the aeroponic system, data acquisition, controlling the equipment, data transmission module, cloud data processing server, social communication platform, and mobile application. Furthermore, in architecture, the data acquisition section refers to some sensor nodes used in the system to establish a data acquisition module. The data acquisition module is placed in the aeroponic system or near the growth chamber to collect the real-time information from selected parameters (temperature, light intensity, humidity, nutrient solution level, atomization quantity, and photos of the growing plants) and transmit the gathered data to the control and management centre. However, the control and management section refers to the central processing unit (CPU) of the system.

The CPU of the system consists of some primary functions such as Arduino and WRTnod protocols, whose work is to store, manage gathered data from collection nodes, process, and then accurately and automatically send to the web server in real time. Thus, the system can help the farmer and grower to monitor and control the smart aeroponic system remotely using the mobile app. In other words, the plant will be able to talk with the farmer through a mobile app that whether the selected parameters are working well or not.

ADVANTAGES OF SENSOR TECHNIQUES IN THE AEROPONIC SYSTEM

The continuously increasing food demands require rapid improvement and development in the food production system. However, to enhance the quality and productivity of the cultivated crop, peoples are moving towards the modern plant cultivation technologies in agriculture. Thus, the aeroponics is one of the rising plant growing technologies in agriculture as a modern-day cultivation technique, where the plant is cultivated in an air environment, and no any soil support is provided. In the aeroponic system, a number of the parameters are required to control for successful plant growth because there is no any growing medium provided to the plant.

For example, if the plant has some sudden stress and the farmer is not present at the site that means the plant will die. Therefore, the proper management of the crop is essential. In the conventional aeroponic system, grower uses his knowledge, skills, and judgment to adjust and maintain the parameters such as EC and pH meter, minimum and maximum temperature, light intensity, and humidity level through several instruments and checks the readings which are labour-intensive and time-consuming task. To deal with the above problems, the aeroponic system can be developed with a wireless sensor and actuator network for monitoring the key parameters at lower labour cost, time, and without any technical knowledge. The wireless sensor and actuator network offer several advantages including faster response to confrontational climatic conditions and better quality control of the crop that produces at a lower labour cost. This advancement in the aeroponic system through wireless sensor network for monitoring growth chamber environment is beneficial. However, the monitoring system also offers a range of information which could be required by plant scientists or grower to provide a greater understanding of how these environmental and nutrient parameters correlate with plant growth. It is now recognised that plant grower can perfectly and easily acquire the skills needed to operate an aeroponic system. It provides the full control of the system, not by constant manual attention from the operator but to a large extent by wireless sensors.

APPLICATION OF ARTIFICIAL INTELLIGENCE TECHNIQUES IN AGRICULTURE

Agriculture is the primitive and ancient application that humans started first after born on earth. It has an extensive history between numerous industries and very intimately linked to the human development on earth. Moreover, in the past, the agriculture sector was labour-intensive, but the next-generation farmers, researchers, and associated organisations proposed and applied new farming methods, technologies, skills, and knowledge in agriculture as a modern era to reduce labour-intensive task. Presently, the technology is considered as a key tool to overcome many challenges and eases the way how people live. In the past, the many problems of agriculture, especially in irrigation water management, crop yield production, environment predication, and decision-making, were decided by many factors.

In addition, the fertilization often decided by the mathematical equations, formulas, or the experiences of the experts. The cultivation is represented by descriptive and causal knowledge, and diagnosis of pests and diseases is represented by uncertain knowledge. Thus, this knowledge and experience are illogically incomplete and imprecise, and the traditional procedures can not handle them. However, artificial intelligence has its superiority. It could be an effective approach for solving complex problems to the levels of experts using imitate experts. The term artificial intelligence (AI) was developed in 1956, as "the science and engineering of making intelligent machines". It is a abroad discipline, which was developed for the interaction of several types of fields such as computer science, information theory, cybernetics, linguistics, neurophysiology, and psychology. The main purpose of the creation of the intelligence techniques is to find the solutions for complex problems and to work, react, and respond like humans. It performed work better than a well-qualified person and brought positive economic and environmental results.

Artificial intelligence (AI) tools have helped to predict the behavior of nonlinear systems and to control variables to improve the operating conditions of a system's environment. A recently published report highlighted that artificial intelligence is emerging as part of the solutions towards improved agricultural productivity. The global artificial intelligence in agriculture is expected to grow at a significant level. It is employed to improve the efficiency of daily tasks in agriculture such as the adoption of robots and drones, crop health monitoring protocols, automated irrigation system, and driverless tractors.

At present, several research studies have been performed by implementing the artificial intelligent techniques in agriculture. Researchers revealed that some of the developed applications for agriculture are expert systems and software, sensors for collecting and transmitting data, and robotic and automation which are adapted from different industries into agriculture, whereas the expert systems and software is the planning process such as strategic or operational; it has benefited substantially, due to the expansion of personal computer and Internet use. The systems are generated through the structured knowledge base and reasoning mechanisms acquired from a human expert but with an enhanced computational power and speed. These systems can demarcate management zones taking in consideration with relevant factors and able to recommend suitable crop rotations, optimal plant density, water requirements, appropriate fertilizer use, diagnosing pests and diseases for crops, and suggesting preventive or curative measures.

Researchers discussed soft computing and applications in agriculture. The study reported that soft computing is the combination of the computing technologies, such as an artificial neural networks (ANNs), fuzzy logic (FL), and genetic algorithms (GAs). These techniques are opposed to the hard computing method which states to a huge set of stochastic and statistical methods. The hard computing provides inaccurate solutions and results of very complex problems through modelling and analysis with a tolerance of imprecision, uncertainty, partial truth, and an approximation. However, the soft computing techniques are used to achieve tractability and robustness. It provides a low-cost solution with a tolerance of imprecision, uncertainty, partial truth, and approximation.

Researchers developed a BP-trained feedforward ANNs to predict nitrogen status in cotton plants based on a data from a ground-based sensing system. Tumbo and team used an on-the-go system for sensing chlorophyll status in corn using BP-trained feedforward ANNs and fiber optic spectrometry to acquire spectral response pattern data in corn fields. Researchers developed a texture-based weed classification method consisting of a low-level Gabor wavelet-based feature extraction algorithm and a high-level ANN-based pattern recognition algorithm.

TYPES OF AEROPONIC SYSTEMS

Low-pressure Aeroponics (LPA)

This is the most commonly used aeroponic type used by most hydroponic hobbyists due to its ease to set up, availability at any hydroponic shop, and low cost.

Low-pressure creates droplet size much different from the high-pressure aeroponic system.

What you need for this system is just like any hydroponic system – a pump strong enough to move the water onto the sprinkler heads to spray water around the plant root zone.

High-pressure Aeroponics (HPA)

This type of Aeroponics is more advanced and quite costly to set up as it would require specialized equipment. So they are often used in the commercial production rather than home growers.

The HPA must run at very high pressure to atomize water into tiny water droplets of 50 microns or less.

This system creates such a fine droplet size that create more oxygen for the root zone than the LPA, making it the most efficient system among all aeroponic types.

Ultrasonic fogger Aeroponics

Ultrasonic fogger Aeroponics, or commonly called fogponics, is another interesting type of Aeroponic system.

As the name means, growers would use an ultrasonic fogger to atomize water into super small droplets of water. These are very tiny and you will see it in the form of fog.

Though plants roots find it easier to absorb water in tiny size, there's little moisture in the fog created, and when running over time, it can easier create the salt that can clog these foggers than other Aeroponic types.

AEROPONICS SYSTEM FOR BEGINNERS

Aeroponics System offers the possibility to enhance crop production and diminish costs compared to traditional farming methods or to any alternate soilless technique of farming. Aeroponics System successfully utilizes every vertical distance for either farming or production of the greenhouse. Every area available in the greenhouse is it horizontal or vertical may be used for the maximum production of food per area.

Difference Between Aeroponics and Hydroponics

Aeroponics System is the method of growing plants or plants with root system vulnerable (exposed) to air. Hydroponics system is your technique of cultivation of plants or plants in nutrient solutions under soil-less conditions.

Hydroponics **Aeroponics**

Fig. 6: Hydroponics Vs Aeroponics (Source by Indoor Farmer).

Importance of Aeroponics System

The aeroponics growing of plants is becoming a household name for growing plants for their daily food at their kitchen table in metropolitan regions. When the prices of vegetables are growing high and the quality is not fresh farm, above all they all are not chemical free. Because this is a soil-less technique of growing plants, the manual work is cut down considerably while leaving no traces of soil stains. Once it is designed and placed, it begins to operate mechanically with minimum garden work is done to be taken care of. Aeroponics farming can be done to develop few plants to a high number of crops and types growing together by utilizing the maximum distance of a given unused space in any part of the home.

- **Increased Air Exposure in Aeroponics System:** Air in the rhizosphere or root zone is important for the plant to grow healthy. Aeroponics System provides the best environment for the source of clean air and nutrient solution delivering oxygen and nutrition to plant roots which stimulates plant growth. More aeration around root zone helps to stop pathogen formation. Plants require 450 ppm to 780 ppm concentrations of Carbon dioxide which is available in the aeroponics system for photosynthesis.
- **Water and Nutrient Supply in Aeroponics System:** In this method, delivery of nutrients and water to plant roots is done by sprayers, foggers, misters, or some other devices that may produce a fine mist. Water droplet size is the trick to root growth. The 20 to 100-micron droplet size generates excessive root hair system for sustained plant development and massive droplets over 100 microns lead to less oxygenation.

- **Diseaseless Environment:** Greenhouse when well-maintained, chances of plant disease and insect infestation are very less. The transmission of disease from plant to plant can be lessened. An infected particular plant could also be isolated and destroyed without infecting or damaging other plants. Since the greenhouse environment is disease free, more aerated, and nutrient rich; crops could be increased at higher density giving maximum production.

Design and Materials Required for Aeroponics System

The major design factor in the aeroponics system is creating the lowest possible water droplet size. A fine mist will promote bushier, bulkier root hair system which will have significantly more surface area thereby nutrient and oxygen absorption will be high. The origin growing chamber must fulfill the following criteria, it has to be light proof, should have good air ventilation, and must hold humidity. Normally, much of the growing chamber immature designs are airtight not allowing plenty of oxygen because the water might spill out. A design that may equilibrium humidity, oxygen, light, and continuous nutrient supply all four at a time is the best design.

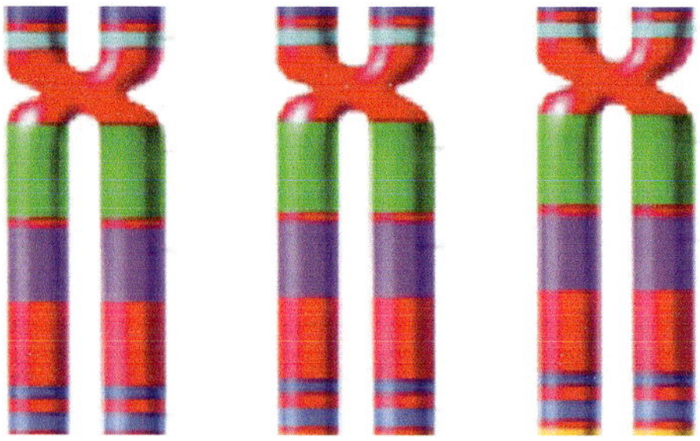

Fig. 7: How Aeroponics Works

Materials to Build a Basic Aeroponics System:

- Container or reservoir or tank to hold the nutrient solution.
- A container is required for root growth.
- A good submersible water pump.
- PVC Pipes and tubes, PVC tee union, PVC elbow union, for water distribution.
- A timer is needed for the water pump.
- Mister or sprinkler heads are required for spraying.

Crops are grown in Aeroponics System or Types of Plants to Grow in Aeroponics System

In the aeroponics kind of farming, plant seeds, cuttings, or saplings are put in a netted cup suspended in a growing room, nutrient-laden water is supplied in mist type supplying nutrients for germination and crops to grow. Even though there is a limitation for a number of kinds of plants to develop, a grower has every opportunity to research in growing more forms of crops. Some types of plants (some need trellises) that are being cultivated successfully are as follows.

Vegetable Plants for Aeroponics System: Cauliflower, Cabbage, Broccoli, Beans, Tomatoes, Cucumbers, Gourds, Potatoes, Beets, Carrots, Eggplants, peppers, Corn Salad, and much more.

Fig. 8: Growing Tomatoes and Lettuce in Aeroponic System

Fruit Plants for Aeroponics System: Melons, and Strawberries.

Leafy Greens for Aeroponics System: Wheat Grass, Spinach, Lettuce, Basil, Amaranth, Chives, and much more.

Herbs for Aeroponics System: Mint, Skullcap, Stinging Nettles, Ginger, and Yerba Mansa.

Flower Plants for Aeroponics System: Nasturtiums, Violas, Dianthus, Marigold, Pansies, Hyacinth Bean, Calendula, and much more.

PROS AND CONS

Pros

Benefits of aeroponics include:

- Maximum nutrient absorption for plant roots due to no growing medium
- Massive plant growth because plant roots are exposed to oxygen 24/7. This promotes healthy and fast-growing plants. The mist used on the roots can also be sterilized to prevent plant diseases.
- Higher yields
- Considerably fewer nutrients and water used on average compared to other systems because of higher nutrient absorption rate. You can help the environment by using less water and human labor
- Mobility. You can move around easily plants or even the whole nurseries as all you need is to move the plant from one collar to another
- Little space required. You don't need a lot of space to be able to set up this system. Plants can be added up one on top of each other. With this type of modular system, you can maximize the use of limited space.

- Easy system maintenance because there is no growing medium used. However, you need to disinfect the root chamber regularly, and periodically the irrigation channels and the reservoir
- System can be cleaned easily
- Easy to replace old plants with new ones
- Great educational value. Adults and kids can use this system to grow pet plants and learn a great deal about plants without needing to get their hands dirty.

Cons

- Besides many great advantages, aeroponics also has the downsides that cannot be overlooked such as:
- Require constant attention with pH and nutrient density ratio because this system is sensitive. Understanding what is the right ratio and applying this concept may be difficult for beginners and can only be attempted by those who are more familiar with such systems.
- The cost for initial set up can be high, which can be many hundreds of dollars each
- Require constant supervision
- Susceptible to power outages. You will have to water your plants manually if this happens.
- Require technical knowledge as it one of the most technical to set up. As there is no soil to absorb the excess nutrients, you need to have a good knowledge about amounts of nutrients required by the plant roots
- Dependence on the system. An aeroponic system is made of mist nozzles, high-pressure pumps, and timer. If one of these breaks down, your plants will die easily.
- Require regular disinfection of the root chamber (a common disinfectant is hydrogen peroxide) to prevent root diseases.
- Microorganism can be introduced to plant roots through water.

9

PLANT MANAGEMENT

Robotics and remote systems have been used for supporting nuclear facilities since the dawn of the atomic age. Early commercial nuclear plants implemented varying levels of automation and remote operation, such as maintenance activities performed on the reactor pressure vessel and steam generators. Over the past several decades, there has been a steady progression toward incorporating more advanced remote operations into nuclear plants to improve their efficiency and safety.

In addition to the main above-mentioned field operations, robots that focus on plant architecture have been included. More specifically, they focus on material handling, pruning/thinning, and string twining. Here, the robots that have been developed for these types of field tasks are presented (Table 1).

Table 1. Plant management agrirobotic systems for various crops.

Crop	Perception	Operation
Apple tree	Force sensor	Pruning
Hop	Camera, wire detecting bar	String twining
Grape	Three	Cameras pruning

Material handling is not a task of high importance in the production cycle, however it is crucial in growing operations such as plant nurseries. Therefore, Harvest Automation Inc. developed an electric, fully autonomous material handling robot for unstructured environments (e.g., greenhouses, hoop houses, etc.).

As has already been indicated, high-value crops get most of the attention, and one of the most studied crops is grapevines. Subsequently, both pruning robots under review focus on grapevines. The first one was developed by a commercial company titled Vision robotics, and therefore no technical information is available, while the second one was developed by the University of Canterbury. In order to effectively prune grapevines, the robot would completely cover the plant to block the sunlight, and then use a six-jointed robot arm mounted with a cutter, and three-color cameras in combination with LED lights. Similarly, a robotic end-effector for the pruning of apple trees resulted in the branch cutting force being in a logarithmic relationship with the branch diameter.

String twining is a labor-intense task in crops such as high trellis hop, and there is a high demand for mechanization of this operation. To this end, a string twining robot has been developed, which achieved a success rate up to 97% while was moving with speed of 0.19 m s-1.

Recently, research in automatic recognition of diseases has been rapidly growing, with potential applications for developing robots able to recognize single plants, locate and identify diseases, and start routines for disease management. This paper aims to provide details of that new generation of robots that could support plant pathologists.

ROBOTIC TECHNOLOGY AND PLANT MANAGEMENT

Agriculture must produce enough food, feed, fiber, fuel, and fine chemicals in next century to meet the needs of a growing population worldwide. Agriculture will face multiple challenges to satisfy these growing human needs while at the same time dealing with the climate change, increased risk for drought and high temperatures, heavy rains, and degradation of arable land and depleting water resources. Plant breeders seek to address these challenges by developing high yielding and stress-tolerance crop varieties adapted to future climate conditions and resistant to new pests and diseases. However, the rate of crop productivity needs to be increased to meet projected future demands. Advances in DNA sequencing and genotyping technologies have relieved a major bottleneck in both marker assisted selection and genomic prediction assisted plant breeding, the determination of genetic information for newly developed plant varieties.

Traditionally plant traits are quantified using manual and destructive sampling methods. These methods are usually labor-intensive, time-consuming, and costly. In addition, manual sampling and analysis protocols generally involve many steps requiring human intervention, with each step increasing the chances of introducing mistakes. Often the plant and its organ is cut at fixed time points or at particular phenological stages in order to measure its phenotypic traits. This method destroys or damages the plant at one time point, disallowing the temporal examination of the traits for individual plants during the growing season. For example, yield measurement (such as plant biomass and grain weight) is invasive and more labor intensive compare to the measurement of plant height and leaf chlorophyll content (measured by a handheld sensor). As a result of the labor and resource intensive nature of plant phenotyping, many plant breeders rely solely on a single measurement most critical to their efforts: yield. However, yield is considered as one of the most weakly inherited phenotypes in crop breeding.

Robotic systems have been playing a more significant role in modern agriculture and considered as an integral part of precision agriculture or digital farming.

We define the robotic technologies as a system having three components: (1) a sensing module that senses the target (plants or crops) and its environment, (2) a computational module to interpret the sensed information and form adaptive (or context-specific) decisions, and (3) an actuation module to complete certain desired operations (e.g., robotic probing, trait measurements, and navigation).

For example, the robot makes decision based on the existing status of environment, obstacles, and plant geometry to manipulate a robotic arm to locate an imaging system with less occlusion and collision free close to plant organs, find appropriate target point on the leaf and control the end-effector based on the leaf angle for effective grasping, or accurately navigate the ground-based vehicles between crop rows. With this definition, systems like LemnaTec's conveyor-based phenotyping platform was not considered in the review, because the plant movement usually follows a pre-defined schedule and no adaptive decision is made during phenotyping. Also not considered in this review are self-propelled ground vehicles or unmanned aerial vehicles that are merely used as a sensor deployment platform with no automated path planning or navigation.

Different artificial intelligence (AI) technologies such as deep learning, fuzzy logic, and genetic algorithms are actively used for control of the phenotyping robots. In recent years, deep learning techniques has gained increased interest to guide robotic manipulators and mobile platforms. In this regard, deep neural networks (DNNs) are commonly used to detect different objects in images such as crop rows, plant organs, soil, and obstacles. DNNs are typically operate directly on raw images and actively learn a variety of filter parameters during the training of a model presented a DNN models to detect the vine trunks as reliable features and landmarks to navigate a mobile robot in a vineyard.

There are three motivations behind writing this review paper. Firstly, robotic technologies in agriculture have seen rapid advancement recently with many emerging applications in plant phenotyping. A timely review of the literature is warranted to summarize the newest development in the field. Secondly, there is large and growing interest from the plant breeding and plant science communities in how these new technologies can be integrated into research and breeding programs to improve phenotyping throughput and capacity. Thirdly, robotic phenotyping has advanced through cross-disciplinary collaborations between engineers and plant scientists.

Outlining capabilities, goals and interests across these two very different disciplines may help readers to identify research gaps and challenges as well as provide insight into the future directions of the plant phenotyping robotic technologies.

The rapid development of new technologies and the changing landscaping of the online world (e.g., the Internet of Things (IoT), Internet of All, cloud-based solutions) provide a unique opportunity for developing automated and robotic systems for urban farming, agriculture, and forestry. Technological advances in machine vision, global positioning systems (GPS), laser technologies, actuators, and mechatronics (embedded computers, micro-sensors, electrical motors, etc.) have enabled the development and implementation of robotic systems and "smart" (intelligent) technologies for precision agriculture.

Machine vision technologies have been applied to agriculture to identify and locate individual plants, with prospective use in highly-automated contexts such as greenhouses. As reported, the object recognition is considered one of the hardest problems in robotics and computer science. In an environment, such as a garden or open field, there is no guarantees that an object has an exact size or shape, due to growth conditions and interaction factors among plants, organisms, and environment. Even the light plays a significant role compared to in a controlled environment; objects such as plants, leaves, weeds, and disease spots might be obscured or show spectral highlights due to the reflection of light. Additionally, machine vision navigation systems have been developed to guide autonomous vehicles in agricultural fields.

Robotic Seeding and Plant Management

Important features for plant management have been implemented in robotic farming, especially for conventional tasks such as tilling, sowing, harvesting of grains, planting, watering, and harvesting of fruit; these activities can be also managed by Farm Management Information Systems. Seed planting represents an important task in plant management; thus, autonomous agriculture robot prototypes have been specifically designed and developed for seed sowing tasks. Researchers developed an agricultural robot in which distances between crop rows and crops represent the only human input.

Plant row detection was achieved utilizing IR sensors. Automation of this activity is particularly important in greenhouses, because the high plant density and short life cycle of many cultivated plants impose frequent and massive planting procedures. Automatic vehicles can use the basic guidance structure provided by the iRobot Create platform to navigate inside a greenhouse and detect specific plant(s).

The robot iPlant provides one arm implemented for watering and a second arm that consists of three main components, a plowing tool, a seeds container, and an excavation tool, to perform seed planting tasks. Robotic fertilization can be carried out using solutions such as the Ladybird robot, which has an electric drivetrain supported by solar panels, and it is able to capture features of surroundings using a Light Detection and Ranging (LIDAR) laser system. A stereo camera creates RGB (Red, Green, Blue) images of the crops, while IR and UV data are collected using a hyperspectral imaging camera (400–900 nm).

This robotic system can evaluate plant status using machine learning algorithms. It is equipped with a spraying system, attached to a Universal Robots UR5 six-axis robot arm, for plant fertilizing. A smaller version of Ladybird was developed as RIPPA (Robot for Intelligent Perception and Precision Application), the production phase of which is underway. Further research has been carried out on robotic fertilizing; researchers designed a robotic manipulator utilizing laser sensors able to detect the content of water and nutrients in the plant and/or soil, as well as to decide the optimal dose of fertilizer for the plant. This site-specific approach can optimize nutrient and water usage/efficiency.

Robotic Harvesting

The harvesting of fruit from woody plants is a frequent challenge for robotic applications as well as for shrub or herbaceous plants. Several harvesting robotic systems have been developed and evaluated for cucumber, strawberry, tomato, asparagus, grain, lettuce, mushroom, and orange crops, among others. Fruit detection and localization is very critical for developing robotic harvest technologies. Researchers provided a review on machine vision for fruit load estimation, suggesting the utilization of integrated color and shape-based feature classification. Most of the supervised learning methods require high numbers and accurate training samples, but deliver better results than simple image processing techniques. Modeling and analyzing 3D plant and tree structures for robotic applications is computationally intensive and a time-consuming procedure.

In general, there are two types of 3D modeling systems: active (e.g., laser systems such as LIDAR—Light Imaging, Detection, And Ranging-, LIDAR-based LemnaTec Scanalyzer HTS—High Throughput Screening-, Microsoft Kinect etc.) and passive sensors (e.g., digital cameras and photogrammetric techniques). The main drawback of the active techniques is the high cost of these systems.

The passive techniques are less expensive than the laser-based active techniques, and can produce more detailed information in addition to faster data acquisition and process rates. However, the critical issue for robotic harvesting is identifying crops that may overlap, or be subject to varying lighting conditions. Recently, tomato and apple detection in highly variable lighting conditions was performed by applying an algorithm on multiple feature images (image fusion).

The application of deep neural networks has also led to the development of new object detectors such as the Faster Region-based Convolutional Neural Networks (Faster R-CNN) that was adapted for fruit detection using color (RGB) and near-infrared (NIR) images. The combination of multi-modal image information via early and late fusion methods leads to an enhanced Faster R-CNN model with higher detection accuracy.

Furthermore, distributed autonomous gardening systems have been utilized for urban/indoor precision agriculture applications, where multi-robot teams are networked with the environment [10]. Post-harvest robotic technologies have been developed based on machine vision for disease detection as well as fruit and product sorting.

Closed or Open Spaces: Challenges for Robotic Plant Management

As researchers suggested, systems for automated plant management may rely on the vision of an Ecology of Physically Embedded Intelligent Systems, or PEIS-Ecology. In this vision, a network of cooperating robot devices embedded in the environment is preferred over having an extremely competent robot in a passive environment. This represents an ecological view of the robot-environment relationship, in which the robot and the environment are seen as parts of the same system, working to achieve a common goal; a vision particularly suitable for greenhouses or well-defined environments such as gardens. In this case, a robot (or PEIS) is intended as any device incorporating some computational and communication resources, and is able to interact with the environment via sensors and/or actuators.

PEIS are connected by a uniform communication model for exchanging information and each component can use functionalities from other PEIS. A model like that could be applied in the agricultural sector as well, particularly in highly controlled conditions which can be obtained in greenhouse cultivation. In a traditional approach, the robot would use its sensors to acquire information from the environment, i.e., to self-localize and recognize/identify plants, evaluate soil moisture, and water the plants. In comparison to the PEIS approach, the robot will recover information from the environment, such as calculate its relative position in the garden or greenhouse (e.g., using cameras), identify plants by RFID (Radio Frequency Identification Device) tags attached or inserted in the plant, monitor soil moisture, and transmit data wirelessly to an automated irrigation system. The lower complexity of robotic devices used in this vision seem to fit well in the agricultural context, where "objects" are stressed by many physical factor (heat and light, soil and sand, water, etc.), which allows for an increase of maintenance services.

Open-field robotic management is a quite difficult task, mainly due to limited control of environmental conditions, increased maintenance workload of high-tech devices, and lack of complete knowledge of objects within the operative range of robots. Thus, robots for open-field application are mainly involved in agronomic tasks or weed control. However, an emerging challenge in precision agriculture is the use of Unmanned Aerial Vehicles (UAVs), for which applications are countless and very promising for plant health monitoring. UAVs can be used for plant disease and abiotic crop disorder detection, such as rice lodging or crop nutrient diagnosis via spectral analysis. UAV systems have also been tested for pathogen detection in crops. Pathogen detection can be achieved by analyzing the crop canopy, or detecting declines in chlorophyll levels (commonly expressed as a reflectance decrease in the near-infrared band). Similarly, for monitoring applications for abiotic stress, this approach seems effective when the pathogens lead to a widespread attack on crops. At this stage of development, this approach should be considered for remote sensing applications, rather than for robotic disease management, where the robots recognize and control the disease. Disease detection and control in open-fields (not only in well-defined environments) could be achieved by utilizing aerial (UAV) and ground robots (Unmanned Ground Vehicles, UGVs). In this study, UAVs were used to collect spectral images and analyze them to identify "critical" regions. This information was transferred wirelessly to the UGV, which navigated to the "critical" areas to analyze and collect leaf samples. The UGV first examined the plant (strawberry) using a spectral camera and, if it was classified as "diseased," it collected leaf samples using a manipulator and web cameras for leaves localization.

ROBOTIC PRECISION PLANT PROTECTION

Abiotic Stress

Among abiotic stresses, water management represents a core topic in robotic management, due to its importance in greenhouses, gardens, and urban greening. These are environments in which robots and automated systems fit well due to the limited extension of surfaces or items to be monitored, environmental conditions that are partially controlled, and the high value of plants. Automated and smart irrigation systems are variable in greenhouses, because growth performance strongly relies on highly efficient water management. Pathogen development is also dependent on water disposal, while a water surplus may lead to increasing costs or cause issues for water discharging in urban greening.

An interesting approach in plant management, with potential effects in protection from abiotic stress, is suggested by symbiotic robot-plant bio-hybrids. Researchers investigated the symbiotic relationship between robots and plants in order to develop plant-robot societies able to create meaningful architectural structures; the robots could cause artificial stimuli in addition to the natural stimuli of the environment. Thus, a robot-plant bio-hybrid could serve as a green wall able to adapt to environmental stimulus, i.e., adapting plant growth and foliage distribution according to solar radiation and temperature, establishing a system that could be used for urban mitigation. Solar access is an important factor and may represent a constraint to urban agriculture, since most edible plants have a relatively high sunlight requirement. Thus, the access to sunlight must be considered.

However, symbiotic plants seem to have the potential to manage multiple requirements of future architectures. Symbiotic plants may also act as bio-sensors, because the bio-hybrid part has set environmental, electrophysiological, photosynthetic, and fluidic sensors. This system monitors the status of plants and their environment. Plants represent a primary, sensitive biological sensor of different environmental parameters, and technological sensors convert the plants' response into digitally represented information, creating a phyto-sensing system . This brings an advance in the current use of bioindicators for air pollution assessment. Moreover, a set of actuators can maintain the plants' homeostasis in terms of watering, chemical additives, spectral light, or temperature, contrasting abiotic stress and minimizing input. These features are very attractive to future urban greening maintenance and protection.

Machine learning techniques can aid in the recognition of abiotic stress, since it is very challenging to distinguish between a variety of diseases with similar symptoms (e.g., biotic or abiotic stress). Self-organizing classifiers (counter-propagation artificial neural networks, supervised Kohonen networks, and XY-fusion) have been able to distinguish healthy and diseased canopy (using hyperspectral images), and to distinguish between nitrogen-stressed and yellow rust-infected wheat.

Biotic Stress

Weed Control

Besides the development of robots for auto-navigation purposes, weed control represents the main field of research, mainly due to (a) the economic relevance of the topic and (b) the feasibility in distinguishing target (weeds) from non-target (cultivated plants) items by machine vision technology. Distinguishing between weeds and cultivated plants represents the first step for achieving effective weed control.

The use of agrochemicals for weeding offers an effective solution to growers, but some concern may arise when considering the environmental impact of this approach. In order to avoid environmental pollution, chemicals could be distributed only towards weeds (target-based applications) or mechanical methods may be applied. Both of these approaches are difficult to implement after seeding, transplanting, or during vegetative growth due to technical difficulties and high costs of the target-based systems. Currently, identification of weeds relies on human capability in discerning (identifying and distinguishing) plants visually. This operation (human supervision of weed control) is time-consuming and expensive.

The "screening for weeds" procedure is similar with the "screening for anomalies" in an industrial production chain, an activity widely managed by automatized machinery. Deploying multiple robots could reduce the cost of weed management in terms of energy, labor, and agrochemical usage. In the RHEA project (Robot Fleets for Agriculture and Forestry Management), robots are integrated in tractor units and equipped with machine vision-based weed and crop row detection systems. This approach seems able to detect up to 90% of weed patches and eliminate most of them by using a combination of physical and chemical means.

However, the applications of multiple robots have to be evaluated considering the cost of the individual units and the issues related to collisions, communication, and safety. Diagnostic real-time tools for weed detection are limited in their commercial availability. For mechanical (target-based) weeding systems, the robotic system needs only to identify and distinguish the cultivated plants; all the other plants (weeds) will be mechanically removed.

Conversely, some issues may be linked to chemical protection because the use of broad spectrum compounds may be necessary, or highly precise spraying systems need to be implemented.

Diagnostic Specificity: The Challenge of Microorganism Control

Categorizing plants according to their morphological parameters is a very difficult task, especially through visual (human) observation. Thus, identifying diseased plants from symptomless ones is an extremely challenging task. The difficulty is considered so intrinsic due to the non-specificity of many symptoms that can be present. Moreover, the degree of alterations of morphological parameters, due to biotic or abiotic stress (such as dimension, shape, or colors), may be very low compared to the variability of appearance in healthy plants; particularly during the first stage of infection.

Control strategies against microorganisms have to be carried out before the occurrence of disease outbreak in order for it to be effective, and diagnostic specificity is mandatory for effective plant protection. Non-destructive methods for plant disease detection have been developed in the few last decades, belonging to two groups: spectroscopic and imaging techniques or volatile organic compound-based techniques. In particular, sensors may assess the optical properties of plants within different regions of the electromagnetic spectrum and they could detect early changes in plant physiology due to biotic stresses that can lead to change in tissue color, leaf shape, transpiration rate, canopy morphology, and plant density.

Thermography, chlorophyll fluorescence, and hyperspectral sensors are the most promising technologies. Plant phenotyping is also critical to understanding plant phenotype behavior in response to genetic and environmental factors. Nevertheless, the use of spectral analysis is a matter of debate in plant pathology, due to similar symptoms or physiologic alteration caused by different diseases. Pathogens detected by image analysis is rapidly increasing, mostly involving widespread pathogens frequently recurrent in arable agriculture systems.

This represents the main field of application because this approach may greatly reduce the workload of routine pest management, but quarantined pests may be also involved by research. In these cases, the requirement for diagnostic tools may be different, because of the need for large-scale monitoring diagnostic techniques (and relative fast and low cost of execution) and high diagnostic specificity, due to the significant workload and social impact that may originate from recognition of positive plants. Researchers developed a recognition method based on visible spectrum image processing to detect symptoms of citrus greening diseases (also named citrus Huanglongbing, HLB, caused by Candidatus Liberibacter spp.) on leaves. The experimental results showed a high detection accuracy of 91.93%. Researchers presented an HLB detection system at the pre-symptomatic stage utilizing polarized imaging. Citrus canker caused by Xanthomonas axonopodis caused foliar symptoms that were analyzed to evaluate the efficacy of image analysis. The image analysis was more accurate than visual raters for various symptom types (lesion numbers, % area necrotic, and % area necrotic and chlorotic), but symptom heterogeneity or coalescence of lesions could lead to discrepancies in results, making full automation of the system not yet affordable.

Researchers developed a vision-based, novel transfer and deep learning technique for detecting symptoms of leaf scorch on leaves of Olea europaea infected by Xylella fastidiosa, with a true positive rate of 95.8 ±8%. These vision-based systems can be mounted on mobile vehicles (e.g., autonomous vehicle) to optimize disease detection procedure and disease management.

Pathological Considerations in Robotic Fruit Recognition

Robotic fruit recognition is one of the most relevant topics for agricultural robot development, due to the economic impact of full automation and mechanization in harvesting. Color represents one of the simplest identifier to distinguish fruit from woods and foliage. Infrared light with wavelengths about 700–1000 nm reflects well by all parts of the plants, while red light (at about 690 nm) is not reflected well by unripe fruit, leaves, and stalks, but is reflected well by red ripe fruit. This feature can be used to recognized even small red fruits, such as cherries, using different light beams. Red and infrared signals were detected separately using lock-in amplifiers, with the aim of eliminating the interference effects caused by environmental light.

Further refinements were necessary due to red light reflections on wood, but the reflection of fruit cause a specula phenomenon that can be used to recognize the center of the cherry. Thus, fruit and obstacles, such as wood or leaves, can be properly recognized to accomplish an efficient harvest. This system should be susceptible to plant disease. Most cherry diseases are related to twigs, branches, or wood, causing a reduction in foliage biomass that may alter light reflection; while diseases such as necrotic ring spot (Necrotic ring spot virus) or leaf spot (Blumeriella jaapii) may reduce the reflection of infrared laser beams due to spots on the leaves. Cherry recognition could be compromised by brown rot (Monilinia fructicola), that could strongly reduce reflections of red laser beams due to browning on ripening fruit and coverage with gray masses of spores.

Even if the biotic stress interferes with the recognition of plant organs, alterations in light reflection parameters should represent an interesting starting point in order to adapt harvesting-based solutions to disease detection ones. Moreover, the use of 3D vision sensors, essential in order to recognize fruit covered by foliage, will be useful even to evaluate symptoms caused by pathogens that are erratically distributed.

Fruit detection via color can be a difficult task when the fruit has the same color as the vegetation, such as in the case of cucumber. The reflection of leaves at 850 and 970 nm is nearly the same. Fruit showed a significantly higher reflection at 850 nm, and hence, the use of different filters allowed the distinction between the reflectance of leaves versus fruit, achieving high efficiency in fruit recognition. It is likely that a leaf disorder, such as fungal, bacterial leaf blight (Alternaria cucumerina, Xanthomonas campestris), or anthracnose (Colletotrichum orbiculare), may cause rupture in the reflection consistency. Rot caused by Rhizoctonia solani may alter ratios in 850–970 nm reflectance, but these disorders could also compromise the whole recognition method, providing useless value for pathology purposes.

Other approaches in fruit and leaf recognition for harvesting purposes, not based on color, could lead to interesting applications on disease detection. Researchers indicated that recognizing plants and tomatoes was a difficult challenge due to the complex geometry of the leaves; the foliage interferes significantly with the lighting in unpredictable ways. Tomatoes are generally round, but the variations in size and shape may be numerous. Lighting may also complicate their detection due to obscuration of other fruit or leaves, and their skin may reflect spectral highlights.

To solve the identification of red or green tomatoes on plants, filter-based object recognition can be carried out. Possible tomato locations are derived from a collective estimate of filters based on shape, color, size, and spectral highlights of tomato skin. Some filters, such as spectral highlights, are not robust and their performance is affected by changing light conditions and by the object's shape, leading to numerous false positive. Two parameters were selected for fruit classification and identification (e.g., red, green tomatoes); color and fruit smoothness. Fruit smoothness was the most critical factor for identification; fruit identification using only color features could lead to false positives (between green tomatoes and leaves). This approach may lead to interesting applications for symptom recognition.

A machine vision technology able to identify fruit or leaves independently from color could individuate the specific part of an image or a plant, on which to apply to other detection tools (i.e., hyperspectral imaging) or further filters, limiting interferences of foliage or background. Moreover, if smoothness and color are the key factors in identifying fruit, breakage in homogeneous areas should be used as symptom alerts. Early blight symptoms caused by Alternaria solani cause sunken areas and a black and velvety appearance, while anthracnose, revealed by small, indented spots on tomato skin caused by Colletotrichum coccodes, results in a decay in tomato flesh. Bacterial speck (Pseudomonas syringae pv. tomato) and bacterial spot (Xanthomonas campestris pv. vesicatoria) are characterized by raised brown specks or pimple-like dots, while bacterial canker (Clavibacter michiganensis subsp. michiganensis) causes raised black and white spots. Alterations in skin smoothness and color may therefore also indicate by physiological disorders such as fruit cracking, catfaced fruit, or sunscalding. The combined spot alterations of smoothness and color caused by these fungi, bacteria, or disorders should be properly recognized within the area that, for the same feature, was identified as fruit.

Thus, the robot could harvest only symptomless tomatoes, reducing post-harvest work. Other biotic stress such as late blight (Phytophtora infestans) that causes mainly color alterations on fruit (i.e., brown blotches), or physiological disorder such as blotchy ripening, should be more difficult to distinguish, due to low impact on smoothness of fruit skin.

Further sensors should be implemented/developed in order to integrate robot potential, such as hyperspectral techniques that provide much information over the 350–2500 nm spectral range, with a resolution of less than 1 nm. These techniques have shown high potential in plant pathology because the information retrieved from hyperspectral image is based on the spatial X- and Y-axes and a spectral Z-axis, generating a detailed and allocated interpretation of the signal object interaction. Researchers were able to identify tomatoes infected by P. infestans, but this was related to canopy disorder instead fruit. Applications to fruit disease are quite rare. Thus, data provided by this class of techniques may be able to integrate performances of robotic fruit recognitions, creating an effective robot pathologist.

PERSPECTIVE APPLICATIONS OF ROBOTIC PHENOTYPING

Phenotyping Robots Has Great Potential to Measure Other Plant Properties

Many Indoor and Outdoor Robots Were Developed to Measure a Wide Range of Plant Traits" introduced the robotic systems for indoor and outdoor applications to measure several different plant traits. However, other leaf/stem characteristics are also reliable indicators to detect the symptoms of biotic/abiotic stresses and monitor the plant health during a growing season. Stomatal conductance, gas exchange, and chlorophyll fluorescence of leaves are indicative of their water status, photosynthesis, and chlorophyll content. Stem sap flow and lodging resistance can provide useful information about plant water use and stem strength.

These aforementioned phenotypic traits are still measured manually. On the other hand, new clip-on sensor system can be presented to measure them automatically. The system includes a custom-designed gripper/clip combined with novel sensor(s). The design of these sensing systems is important since the accuracy and robustness of trait prediction models depend on the phenotypic data quality. The design of the gripper and DOF of the robotic manipulator should allow a good and gentle contact between the sensing unit and the leaf/stem.

Sometimes a vacuum mechanism attached to a soft gripper can hold the leaf/stem and help the sensing unit for effective contact and collect accurate data with less damage to the plant organs. Moreover, autonomous robots should gather data with minimum error (high signal to noise ratio). Therefore, sensors with high signal to noise ratio should be selected and accurately calibrated. In addition to the accuracy, the robots should rapidly (short execution time) accomplish their missions. Deep reinforcement learning (DRL) technique is an accurate and reliable method to find an optimal path with nearest and collision avoidance route. This technique can be adopted by phenotyping robots to manipulate a robotic arm for grasping process or to navigate a mobile robot between crop rows.

Although the robotic phenotyping is mainly focusing on leaf and stem, it can be utilized for other plant organs such as inflorescences (spike, panicle, and tassel), flowers, fruits, and roots. The morphometric parameters of inflorescence are highly correlated with yield and grain quality.

Several studies discussed about using image-based techniques (2D images/3D reconstruction) to extract architectural traits such as length and width of inflorescence, inflorescence volume (weight), grain shape and size, grain angle, and number of grains, and number of flowers. In such applications to measure the morphological traits, a robot with LIDAR/camera can be useful to automatically take images/point cloud data from different views of the inflorescence. The physiological traits are indicator for stress or disease. For instance, the temperature of the spikes was used for detecting the plant under the water stress. Conceivably, a robotic arm equipped with a temperature sensor can grasp the spike and insert the sensor into spikelets to record their temperature.

Several properties of fruits such as water content, sugar content, chlorophyll, carotenoid, soluble solid, acidity, and firmness are measured for fruit quality assessment. The spectroscopy/spectral imagery are non-destructive and high-throughput methods to estimate these qualitative parameters.

However, a robotic system can be presented to monitor the dynamics of these attributes for hundreds of growing fruits per day. For example, a portable spectrometer can be attached to the robot's end-effector. After detecting the fruit on the plant, the robot can grasp the fruit and gather its spectral data to further infer its quality parameters. Since the root has functional roles in resource acquisition, the characteristics of root provide valuable information about plant physiological and ecosystem functioning.

In traditional root phenotyping, two different methods are used to acquire images from root (in the soil or soil-free or transparent media). In first method, a camera is mounted on a tripod and moved by a human around the root, and in the second method camera(s)/sensor(s) are set in fixed point(s) and root (plant) is rotated. This is a tedious task and some root information (such as fine branches) might be lost due to less flexibility of the system to take up close images from the complex architecture of root.

Consequently, automated root phenotyping systems can facilitate and improve the traditional root phenotyping in terms of efficiency and effectiveness with acquiring fast and precise measurements. Here, the "plant to sensor" system can be used to examine vast number of roots (or plants) without the need of huge space of greenhouse facility. In this system, the root (or plant) is moved toward a robotic manipulator (equipped with camera/sensor) and located on a rotation table. In each step angle of the table, the root is rotated and stopped in front of the robotic system. Then, the robotic manipulator moves the camera around the root and gather close proximity data from different views (positions and angles). Therefore, more detailed information of root can be captured due to high resolution sensing offered by the robotic system.

ROBOTS IN GREENHOUSES COMPLEMENT THE IMAGE-BASED PHENOTYPING

Automatic greenhouses such as LemnaTec (LemnaTec GmbH, Aachen, Germany) monitor plants using image-based technique. While it has shown great potential to measure and predict the plant traits, many hurdles cannot be handled by this technology. It needs to efficiently manage "big data" problems and also postprocess images to characterize the plant traits. Moreover, this approach is not sufficient for early detection of stress/disease with internal symptoms. Furthermore, this method requires direct measurements using sensors to calibrate and validate of its models to extract the phenotypic traits from images.

Hence, several robotic arms with different sensors can be integrated to the greenhouse for real-time and direct measurement of the chemical/physiological traits. Basically, plants are transported by an automatic conveyor belt and stopped in front of each robotic system. Then, the system uses "sensor-to-plant" concept in which the robot moves toward the plant to take measurements before sending it through the imaging chambers. These stationary robotic systems are designed to operate in indoor environment. Moreover, several robots can be presented to collect data from a specific plant. It is difficult to develop a general prototype that are broadly applicable for different conditions. However, the software and hardware of the robots should be adapted to other species and field-phenotyping applications. The challenge for both type of robots (indoor/outdoor) would be continuously collect and save large amount of data.

Swarm Robot is a New Frontier to Efficiently Accomplish Complex Phenotyping Tasks

Swarm robotics is a new frontier technology which has potential application for proximal sensing of plants, and data/sample collection in a large field. A swarm robotics system composed of large numbers of autonomous robots that are coordinated with local sensing and communication, and a decentralized control system. The application of swarm robots has some advantages which is suitable for large scale tasks. Since swarm robotics has large population size, the tasks can be decomposed using parallelism and can be completed efficiently and consequently it would save time significantly.

Moreover, the swarm robots can achieve the distributed sensing that means they can have a wide range of sensing in different places at the same time.

ELECTRICAL ROBOT FOR AUTONOMOUS SPRAYING IN AGRICULTURE

Precision spraying in agriculture represents an innovative methodology that can help farmers to spray only where needed, using the right amount of chemicals. This can improve the efficiency of treatments, reducing costs and chemical waste. Moreover, precision spraying can help to fulfil legal requirements, in terms of worker health, environmental sustainability, and food safety and quality. However, most suppliers do not offer robots and integrated systems that are able to operate in narrow and steep spaces or in the presence of obstacles, as in the case of greenhouses or terraced cultivations.

The subject of this work is a system capable of distributing inputs (e.g., crop/plant protection products, bio-agro drugs, pesticides, powders) in an efficient, sustainable, and safe way, in environments where standard agricultural machineries (automated or autonomous) cannot operate: greenhouses, mountainous areas, and heroic cultivations.

The adoption of autonomous robots can help in optimizing the agricultural operations, increasing productivity and safety. Moreover, the costs of technological devices (e.g., electrical motors, batteries, computers, sensors) is reduced from year to year, allowing these robotic systems to become easily wide-spread of among farmers. Here common robotic architectures, components, and applications are discussed in depth. Many research activities deal with the adoption of robotics in agriculture. Researchers provide a general description of the potentiality of human–robot interaction (HRI) in many agriculture activities, such us harvesting, handling, and transporting.

Agricultural robots are typically autonomous or semi-autonomous systems that can be operated in several stages of the process. In particular, robots that must operate inside greenhouses and vineyards have to deal with very tight passages while avoiding damages to the plants. Moreover, they should have very good performance regarding path following, during turn maneuvers at headlands. Several localization and navigation solutions have been implemented. A laser-guided path follower technique was tested. The system is stated to be very reliable, but requires several lasers to be mounted in the proximity of corridors. Different approaches were used regarding sensors and algorithms. Some trials were performed using map-based techniques and using motor encoders for localization.

A combination of the Hector Simultaneous Localization and Mapping (SLAM) and an artificial potential field (APF) controller is used to estimate the robot's position and to perform autonomous navigation inside the greenhouse.

However, in most cases, the developed prototypes presented above are just proof-of-concept, and are not sufficiently robust for the real agricultural environment. Our system attempts to bridge the gap between all the prototype solutions described above and those big machines that already work in large crops and farms, which cannot be considered as solutions applicable in greenhouses or heroic cultivations and vineyards. Moreover, in terms of versatility, compared to some of the previous solutions, our system does not require the installation of rails between the rows and can be used in different kinds of crops and terrains.

Figure 1: The developed spraying robot while operating in a greenhouse.

The heart of the system we developed is represented by an agile autonomous obotic vehicle able to move in environments where standard agricultural machineries (automated or autonomous) cannot operate: greenhouses, mountainous areas, and heroic cultivations. The vehicle, shown in Figure 1, is equipped with a "smart spraying system" capable of optimizing the spraying operations.

Furthermore, the developed human–machine interface also allows unskilled operators to plan spraying treatments, to monitor the mission from a safe distance, and to generate a report of the treatments at the end of the operations.

Mobile Tracked Robot

The robot is based on the "U-Go" robot—one of the multifunctional vehicles developed in our DIEEI Robotic Laboratories at University of Catania, mainly to solve problems like transportation, navigation, and inspection in very harsh outdoor environments. It is a rubber-tracked vehicle equipped with electric traction: two motor drivers, connected to two 1 kW 24 V DC brushed motors, acting on the tracks, can be commanded via CAN bus. The compact external dimensions (880 mm (L) × 1020 mm (W)) also make the machine highly maneuverable between narrow passages. It allows the transport of a payload of about 200 kg, at a maximum speed of 3 km/h. It is able to exceed a maximum slope of 30 degrees, and has an autonomy of 6–8 h.

The vehicle is equipped with several sensors for autonomous navigation in rough outdoor environments: a stereo camera, a RTK-DGPS GNSS receiver, an attitude and heading reference system (AHRS), a laser scanner rangefinder, ultrasonic sensors, and motor encoders. An embedded acquisition and control system, based on the sbRIO-9626 board by National Instruments, acts as low-level interface with the sensors and actuators. A custom protocol based on User Datagram Protocol (UDP) is adopted to communicate and interact with the sbRIO over a wired Ethernet or WiFi connection. It is possible to retrieve all the sensors' data (GPS, attitude, etc.) and to send commands to the board. For example, linear and angular desired speeds can be passed via UDP to the sbRIO; this uses the robot kinematics equations to compute the reference signals to be assigned to the vehicle's two motor drivers. This kind of architecture ensures hardware and software modularity: an on-board companion computer or a remote base station can be used to interface and communicate with the mobile tracked vehicle in order to implement high-level control algorithms, while demanding the low-level tasks of the sbRIO.

Localization and Navigation System

This subsystem fuses the information coming from the on-board sensors to accurately determine the position and attitude of the vehicle, used to autonomously execute the assigned trajectories and avoid obstacles.

A Raspberry Pi 3 (RPi) was used as an on-board companion computer, running the high-level algorithms of the localization and navigation system. The two software subsystems (Localization and Navigation modules) run in the RPi as two ROS (robot operating system) nodes. RPi communicates with the sbRIO by means of the aforementioned UDP protocol.

In particular, the Localization node implements an extended Kalman filter to fuse the data coming from the sensors, in order to evaluate robot position and heading. Moreover, the sensor suite and the Localization node allow several high-level tasks to be performed:

- The stereo camera and laser range finder can be used to perform terrain reconstruction and drivable-surface detection; they can also be used to detect obstacles and to perform odometry-free dead-reckoning.
- The ultrasonic sensors and/or laser scanner can be used both to detect obstacles and to localize the robot inside greenhouse corridors.
- However, in the case of study reported in this work, the Localization node uses only a subset of the sensors suite to evaluate robot position and heading:
- Absolute positions are measured integrating the measures of the encoders with the information coming from a Leica Viva GS10, which is a high-precision GNSS receiver; it uses Leica Geosystems SmartNet—a GNSS Network Real-time Kinematic (RTK) correction service delivered over cellular network—to achieve centimeter-accurate positioning without the need for an RTK base station.
- Attitude is derived from X-Sens MTi-30, a MEMS based AHRS.
- A SICK LMS 200 laser scanner, mounted in the upper-front part of the vehicle, is used to detect static and dynamic obstacles.

The information coming from the Localization node is passed via ROS to the Navigation node that allows the vehicle to autonomously execute an assigned trajectory, avoiding the obstacles. Processing the laser scanner data, the potential field method (PFM) is implemented to follow the desired trajectory, assigned as a list of waypoints (WPs) avoiding obstacles: the next desired waypoint attracts the vehicle, while obstacles exert repulsive force. The implemented control algorithms translate the resulting force in linear and angular speeds that are passed to the sbRIO via UDP.

Smart Spraying System

The smart spraying system is based on a standard spraying machine made "smart" by add-on technologies enabling the automatic control of the system. The sprayer unit is composed of a hydraulic subsystem and an electronic control unit.

The hydraulic part of the sprayer is composed of a 130 L tank with a 5 L hand-wash separated reservoir, an electrical (for safe operations inside greenhouses) actuated pump with a 1 kW DC motor with a gearbox, a manual pressure regulator with pressure indicator, an electric flux regulator valve, a pressure sensor, a flow rate meter, and two electric on/off valves.

These two valves can supply, with the chemical fluid, two vertical stainless-steel bars equipped with anti-drop high-pressure nozzles for spraying operations inside a standard tomato greenhouse. The two bars are respectively placed on the left and right sides of the sprayer, and can be operated separately. The pump has a nominal maximum pressure of 25 bar and a maximum flow rate of 20 L/min. Furthermore, the system is equipped with wind speed and temperature/humidity sensors to alert if weather conditions are inadequate for the treatment.

The electronic control unit is composed of one digital I/O module and one analog I/O module with Ethernet interface. A ModBus-TCP-based protocol allows for interactions with the modules from the on-board computer as well as from any other remote station with a wireless connection: by using a suitable commands set, it is possible to operate the hydraulic components (i.e., valves, flow regulator, pump) and to read information from the pressure sensor and the flow rate sensor, as well as other control unit parameters (e.g., main breaker status, pump current, flux regulator valve status).

A dedicated ROS node, running on the RPi, is devoted to the management of the smart sprayer. This node interacts with the modules of the sprayer control unit via the ROS2IP node, which translates ModBus-TCP messages to ROS messages and vice versa. While communicating with the Navigation node and Localization node, the Sprayer Node controls the sprayer, relying on the spraying treatments previously planned by the user or scheduled from a spraying map. While moving, the robot automatically activates the requested valves at the correct waypoint. This aim of this work was not to determine how to obtain the spraying map; nevertheless, when the spraying map is available, the system can use it and locally sprays the required quantity.

Human–Machine Interface (HMI)

The HMI, running on a tablet PC, allows operators with no engineering knowledge to manage the system. The interface communicates with both the sbRIO and the Raspberry Pi using a WiFi connection.

The HMI can be used for four main operations:

- To visualize the position of the vehicle on a map and to monitor the status of the system (battery level, liquid level in the tank, state of the valves).
- To plan the mission. This operation can be scheduled on the basis of a spraying map from which WP and required chemical dose can be processed. It is also possible to select the WPs by clicking on a georeferenced map, if available. The trajectories can also be assigned by using a teach-and-repeat strategy, driving the robot through the corridors along the desired trajectory: the WPs are automatically stored in the tablet PC. In this case, a joystick connected to the tablet PC allows the operator to manually guide the robot. At the end of the procedure, WPs are uploaded to the Navigation node and stored on-board the RPi.
- To plan the spraying treatments. For each waypoint, the operator can define whether to activate the right or left sprayers and the quantity of sprayed product. This info is sent to the Sprayer Node of the Raspberry Pi and stored on-board the RPi.
- To generate a report. At the end of the spraying mission, a logbook is automatically generated on-board the RPi, according to the national legislation implementing the European directives. The log contains personal data concerning the company, the name of the treated crop and its related extension, date of the treatment, and the type and quantity of chemical products. It is possible to retrieve this data from the HMI to generate a report.
- The performance of the spraying system was also preliminary evaluated. Standard procedures using water-sensitive papers (WSPs) were used to check for uniform droplets distribution along the robot path and at corridor ends. WSPs were placed at three different heights: on top of the plants (h), at 0.5 h and at 0.3 h. The chosen positions were at the beginning of the row, at the end of the row, and in the middle point of the row. Note that it is important to check the distribution at the beginning and at the end of the rows, since the robot may overspray the same plant during turns if the flow is not adequately controlled.
- Preliminary analysis of WSPs, showed a good performance of the system. Full numerical analysis of the WSPs has not yet been performed because the main goal of the present work was related to the robotic infrastructure of the whole smart sprayer: electrical tracked vehicle, related mechanical parts and power electronics, hydraulics, and electronics. The system has been labelled as a "precise sprayer" due to its hardware capabilities and the availability of a full set of actuators and sensors able to distribute a known quantity of chemicals.

10

AUTOMATED WEED MANAGEMENT AND CONTROL: A NEW PARADIGM IN AGRICULTURAL AUTOMATION

Weeds are a curse to agricultural production and lead to diminished crop yield and quality. Weed management and control are thus essential to the production of high-yielding and high-quality crops. Advancements in weed control technology have had a huge impact on agricultural productivity. Meeting the food and fiber demands of the world's growing population will only be possible with highly productive agricultural systems in which weed management is a critical component

Machines are created to reduce the human drudgery to certain extent in day to day life. With the advancement in technology, robots are quite competent enough to do the tedious, complicated and repetitive work in any field of science and technology. In some robots, where artificial intelligence is used can perform the task faultlessly in an automated way. In agriculture, the use of robot requires a multidisciplinary approach, where engineering technology must be integrated with the principles of agriculture to make the system sustainable in a long run. Among the various issues in the field of agriculture, the infestation of weed is the major barrier both in crop and non-crop land areas. When the weeds are found in the crop land situation, there it competes with the beneficial plant for various resources such as moisture, nutrient, space, sunlight and CO2. However, in non-cropped areas, it reduces the utilization of land and water. Robotics as a part of mechanical approach may be use to reduce the weed infestation either alone or in an integration with other feasible methods of weed management. The fundamental principle of the use of robotics in weed management is to destroy the weeds, those which are present in inter-row as well as intra-row of crop plants. It is suitable for the wider spaced crops, line sown crops, plantation crops and high value vegetables. Different types of robots may be used for weed control depending upon the need as well as the utility in specific crop land situation. In robotics, the sensor may detect the weed and control the weed chemically or mechanically. It can be used in adverse climatic and soil conditions for weed control in different situations. The use of robotics to control the weed is a technology driven process, which requires enormous automation in farm areas. In the context of Indian agriculture, the application of robotics is in the juvenile stage. Thus, it may be encouraged as a viable and sustainable option to the farmers for their greater benefits in future.

Automated weed control is one of the most important issues related to sustainable vegetable crop production, with the ultimate aim of meeting future demands on global food security. Maintaining weed-free zones around vegetable plants is critically important to crop productivity so that the crop plants do not suffer from weed competition, i.e. for nutrients, sunlight and water. A big challenge to agricultural row crop production is the development of an automatic mechatronic system for non-chemical weed control due to inadequate labour for hand weeding, requirements to reduce environmental impacts associated with pesticides, and a need for more efficient organic production systems.

Agricultural robots have great potential to deliver weed control technologies that are much more adaptable even down to the plant scale. They potentially could direct chemical or cultivation tools to directly target weed plants. Agricultural robots can have these characteristics because they bring recent advances in artificial intelligence (AI) to bear on the control of weeds in crop fields. However, bringing AI and robotics technology to weed control has several challenges that may limit robotic weed control robustness, at least with the current state of technology.

Weeds are plants that are misplaced or undesirable to the purpose of the crop plants in the field. Crop plants are being cultivated because of the economic value to the producer. Thus, any plant can be a weed, like volunteer corn growing in a soybean field, if it is not serving the purposes of the producer's management scheme. Thus plants are weeds based on their location and competition relative to the crop plants.

So robotic weed control is an ill-posed problem until the agricultural producer's intentions for a field are made known to the robot, which will then identify and make decisions about which plants are the weeds that need to be controlled.

Another challenge is that while crop plants are mechanically planted in a structured manner that is compatible with agricultural machinery, weed plants emerge and grow in patterns that are consistent with their ecology. Thus weed plants exist in random patterns in a field. The species of plants that make up the collection of weeds in the field varies. This variability changes at different scales as well. Weed plant emergence changes on a meter-by-meter basis,

Weed management strategies take this variability into account and employ weed control techniques that are general and robust enough to be efficacious to control the weeds. In the development of robotic weed control methods, there are obvious challenges to success. These challenges include informing the robot which plants need to be controlled and determining the distinguishing features of those plants. To control weed plants, their growth needs to be retarded or stopped, while not injuring the nearby crop plants. Implementing the control is particularly difficult when the weeds are in close proximity to the crop plants

This chapter will focus on key work in the development of robotic weeders including weed perception systems and weed control mechanisms. An example of an automated weeding system will be described, and readers will be pointed to where they can learn more.

ADDRESSING THE CHALLENGES OF ROBOTIC WEED CONTROL

Much work has been done to address the challenges of robotic weed control. In the subsections below, we will review the literature to provide a perspective on how the challenges have been addressed. The first section will focus on perception systems which can detect and classify weed plants from crop plants.

Weed and crop plant perception

After decades of research and development, numerous methods have been developed for weed and crop plant perception. The main challenges in crop and weed plant perception are vegetation detection, classification of weed plants and crop plants, and plant localization. Machine vision is the most widely used technique. Studies vary in terms of crop or weed species, complexity of visual scene (from indoor-controlled environments to commercial fields) and the sensors used. Perception methods can be categorized in terms of the vehicle platforms carrying the sensors, and the plant features used in processing.

Satellite-, aerial- (including unmanned aerial vehicles, UAVs), and groundbased (with unmanned ground vehicles (UGV), such as field robots, or commercially available off-highway vehicles) vehicle platforms are common for detecting and monitoring plants. Satellite- and aerial-based sensing were commonly used for large-scale field monitoring in applications such as variablerate herbicide spraying. These platforms have a lower spatial resolution, and the working time is affected by the weather and air conditions.

Ground vehicle-based sensing and low-altitude aerial-based sensing can acquire higher spatial resolution plant imagery enabling accurate detection of crop rows and plant localization for applications such as real-time, in-row weed control. Ground vehicle–based methods, however, must meet requirements such as having clearance over the crop, matching the crop row spacing and being able to traverse the field under a range of soil conditions

Spectral reflectance and biological morphology characteristics were two categories of features commonly used in these methods. In the subsections below, weed plant perception methods are reviewed in terms of the type of features used.

Systems and materials

The automated weeding robot consists of an imaging chamber, a pneumatic123 powered mechanical weed-cutting knife, an on-board computer, and a metal frame to hold all the components to mount the weed control system to the tractor. The imaging chamber includes an electrically-controlled, 2 MPixel-resolution, area-scan, single lens, digital, colour camera (Model Scout scA1600gc, Basler Inc., Ahrensburg, Germany), six first-surface mirrors (Model 0.485 Thickness in Glass Sheets, Kaleidoscopes Inc., Iowa, USA), two sets of six high power UV LEDs (ultraviolet light129 emitting diodes, Model LZ4-00UA00 Ultraviolet 410 nm, 10 Watt, LED Engin Inc., San Jose, California, USA) with LED light reflectors (Model C10437 Boomerang Hexagonal, Ledil Oy, Finland). Figure 1 shows the mechanical structure of the weed knife control system.

Weed knife system design

To develop an automatic weed control system for vegetable crops, a mechanical weed knife was designed at University of California Davis. It uses a pair of mechanical weed knives similar in concept to the one in the commercial vegetable crop thinner reported and the GPS-based weed control system for transplanted row crops, but altered for precision intra-row weed control and crop signalling. Figure 2 shows a front view of the weed knife system where the knife is in the close-state (knife tips are touching). Each knife blade (red parts in Fig. 2) was made from a 6.4 mm thick plate of hardened tool steel (Model Aristocrat D-2, air hardened to Rockwell 60, Precision Marshall Steel, Washington, Pennsylvania, USA) and was shaped into an isosceles triangle, with a triangle base width of 7 cm and a triangle height of 3.2 cm. The knife cutting edge was created by sharpening the two forward pointing sides of the triangular plate.

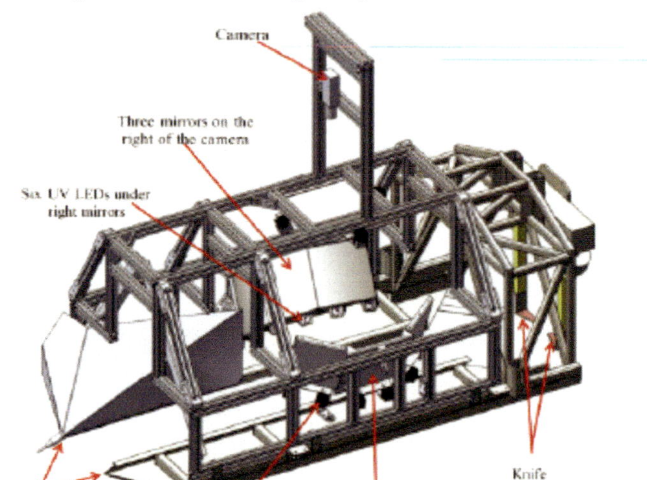

Fig. 1: Mechanical structure of UC Davis weed knife control system, including a amera mounted on top, six first-surface mirrors, twelve UV LEDs, and a pneumatic-powered mechanical weed-cutting knife.

Fig. 2: Front view of the intra-row weed knife where the knife is closed (knife tips touching).

Two arms in a linkage (yellow parts in Fig. 2) were used to fasten the knife blades at their bottom, with the triangular blade tip and sharp-ened cutting edges confronting the forward travel direction. The plane of knife blades was held to be parallel to the ground and approximately 2 cm below the soil surface. However, the knife blade might not always be parallel to the ground, especially when the robot traverses a highly uneven surface. There is some published research on the navigation of robots on uneven terrain. Although, in this work, the relationship between the terrain surface and the robots or the actuators was neglected. It was assumed that the knife blades are always parallel to the surface.

The linkage was designed to support a maximum tomato plant height of 30 cm. Actuation of the knife was controlled by a pair of double acting pneumatic cylinders (Model CCD15-SBP004, Ingersoll Rand plc, Dublin, Ireland) (coloured silver parts in Fig. 2), attached between the knife linkage arms and the frame of imaging chamber on the cultivation sled. An electronically actuated solenoid air control valve (Model A212SD024-D, Ingersoll Rand plc, Dublin, Ireland) was used to control knife motion with air pressure applied to the pneumatic cylinders.

The plant safety zone was set to 10 cm. Third, the knife closed when the plant safety zone had been traversed. Plant spacing (i.e. distance between two adjacent plants) could be defined as a constant or automatically estimated while traveling. For the experiments in this study, it was set to 30 cm for tomato and 13 cm for lettuce fields.

UV imagining system

The UV imaging chamber was designed to block ambient sunlight and capture images under UV illumination to allow the position of main stems of crop plants to be determined based on UV excitation of signalling compounds on the plant labels. The imaging chamber successfully controlled the illumination when the chamber and support skids rested upon the soil surface, so that minimal sunlight leaked in from outside.

An RGB camera (Model piA2400-12gc, Basler Inc., Ahrensburg, Germany) having resolution of 1624 x 1230 pixels, was equipped with a fixed focal length lens (Model Computar M0814-MP2 8mm 1:2.4 2/3", CBC Group Inc., Tokyo, Japan) and was positioned at a proper height above the ground to view plants and all the mirrors. Two sets of three mirrors were mounted on the left and right sides of the chamber, where each set had three 15 x 10 cm mirrors positioned to allow three different view angles surrounding the target plant.

Figure 3 shows the layout of the top view showing the six mirrors. Each mirror in the pair of top-left and bottom-right mirrors (represented by the yellow dotted lines), middle-left and middle-right mirrors (represented by the red solid lines), and bottom-left and top-right mirrors (represented by the green dashed lines) was set up to be parallel to the other in the same pair. The mirrors were used to estimate the 3D geometric appearance of the target plant labels. Twelve UV LEDs were mounted underneath the mirrors and used to illuminate the plant labels from the sides. Recently, different systems have utilised UV excitation light as a useful tool for fluorescence imaging in cotton foreign matter detection and in cellular level biology.

A high intensity, controlled illumination system was developed to be capable of activating the unique fluorescent and reflectance characteristics of the plant signals or labelling. The illumination system design was similar to the existing and successful outdoor machine vision systems for chlorophyll fluorescence but was adapted to the signalling requirements of this work. The intensity of the LEDs could be controlled by an adjustable power supply.

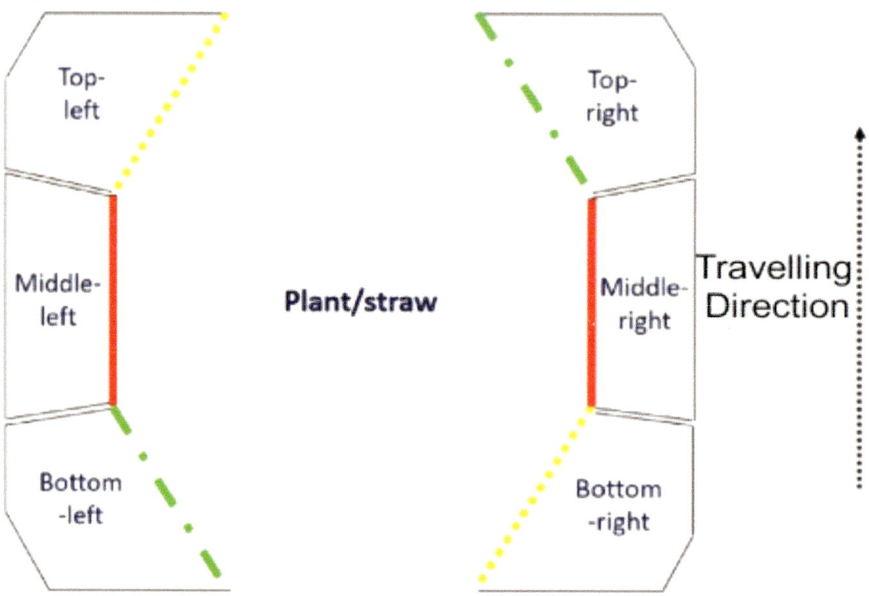

Figure 3: Top view layout showing the six mirrors used to provide side views of the target plant/straw. Top-left and bottom-right mirrors are set up parallel to each other, represented by the yellow dotted lines. Middle-left and middle-right mirrors are set up parallel to each other, represented by the red solid lines. Bottom-left and top-right mirrors are set up parallel to each other, represented by the green dashed lines.

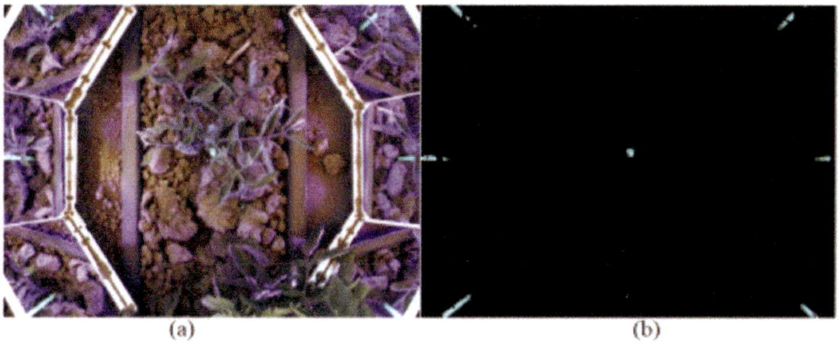

Fig. 4: Image of a tomato plant with a green plant label taken (a) under a mixture of some direct sunlight (for visualisation) and UV light, and (b) under UV light only.

Figure 4 shows an example image of a tomato plant and its green label, captured in the chamber under UV light with some added sunlight (Fig. 4a) for colour-based weed detection and visualisation purposes, and under UV light only (Fig. 4b) for fluorescent signalling.

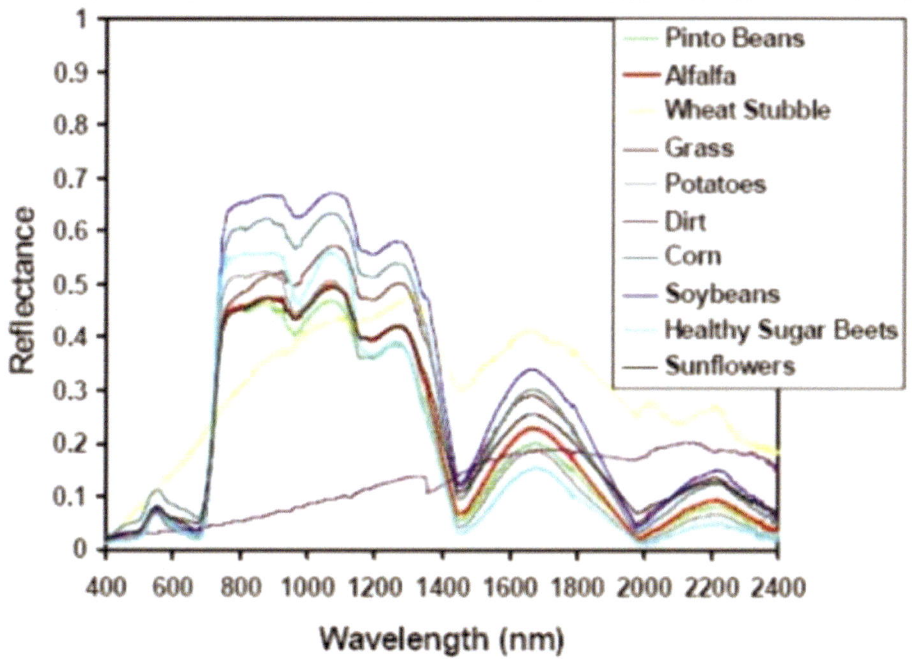

Fig. 5. Spectral signatures of different plants and soil.

Spectral reflectance characteristics

Spectral reflectance characteristics were first investigated, based on the observation that the soils and plants have different spectral reflectance properties. Soil reflectance is typically low across the visible and near-infrared (NIR) regions with a gradually increasing slope going from visible to NIR and IR regions (Fig. 5). There will be some variations in the reflectance spectrum across soil types. Vegetation reflectance has a very distinctive shape with low reflectance in the visible spectrum, but with increased reflectance in the green region. At the red region, entering into the NIR region, there is a dramatic increase in reflectance up into the 50% range.

In general, one of the greatest potential advantages of the reflectancebased techniques introduced above is the robustness to partial occlusion compared to methods analyzing shapes. Additionally, these methods are normally less computationally intensive than shape-based methods. However, the greatest limitation of the reflectance-based methods is the shortage of reliable information. Reflectance information acquired under imperfectly controlled illumination (affected by the sunlight, especially) is subject to change or may even cause algorithms to fail with many factors, such as weather (e.g. sunny or cloudy) and time (e.g. morning and noon-time). Research is still needed to investigate the stability of classifiers using reflectance information in different field conditions and illumination conditions. At the same time, introducing plant morphological characteristics has more potential to improve plant classification performance.

Biological morphology characteristics

Another category of weed plant perception methods focuses on biological morphology characteristics, such as the shapes of plant canopy and leaves in the projected two-dimensional (2D) image plane or in three-dimensional (3D) space, to discriminate plants of different species. These methods were mostly designed for applications such as selective spraying and intra-row weeding, in which crop or weeds need to be accurately detected and localized. Groundand low-altitude aerial-based sensing are two main sensing platforms, on which adequate image spatial resolution can be provided. Techniques such as machine learning are applied to the extracted features to classify species of plants. Since the shapes of plant canopies and leaves are complex and varied, the most challenging task is the investigation of effective and robust descriptors to differentiate different crop species in images.

Historically, most computer vision–based weed perception systems have used light reflectance images as input. Segmentation is usually performed at the beginning to extract vegetation pixels in images. After segmentation, features representing the morphology of plant leaves and plant canopies for plant discrimination are extracted. Common morphological features include length, width, perimeter dimensions, roundness, circularity and convexity of plant leaves or plant canopy. They were reported effective in crop/weed identification. General image feature descriptors, such as features from accelerated segment test (FAST), scale-invariant feature transform (SIFT) features, histogram of gradient (HOG), local binary pattern (LBP), and Gabor wavelet transformation, were also found effective in plant detection and discrimination, and robust to illumination variations.

However, using light reflectance sensors alone to obtain high discrimination accuracy is challenging unless the light is controlled and the plants are sparse. Since most spectral cameras are passive receivers of reflected light, they are dependent on the quality of the reflected light received. The spectral reflectance similarity of vegetation pixels can lead to difficulties in separating leaves or plants with occlusions, and shadow effects or saturation effects caused by uncontrolled illumination may affect the segmentation and feature extraction performance. There is also loss of structural information because the image scene is projected onto a 2D image plane. Incorrect plant identification results may be obtained using features extracted with such conditions.

3D shape features extracted from 3D point clouds were found promising in addressing some of the problems in plant identification associated with reflectance-based sensors alone. These features are more robust to external illuminance changes and shadow effects than those extracted from color images. Plant height, which is a discriminating parameter between crop and weeds at early crop growth stages, was found effective and beneficial in crop/ weed segmentation and classification.

Plant label marking system

The plant labels selected for this experiment were made of 24 cm tall drinking straws made of renewable biomass (model EP-ST910, Eco-Products, Boulder, CO, USA). Three sets of straws were painted with one of orange, green, or pink crop signalling compounds (water-based fluorescent paint model M1800, 17 oz. Athletic Field Striping Paint, Rust-Oleum®, Vernon Hills, IL). Each plant label was then placed in the cell of a vegetable transplant tray and kept together in the greenhouse with each seedling while the roots grew around the base of the straw capturing it in the plug. At transplanting, each seedling and the accompanying plant label (painted straw) were planted together in the field using a standard vegetable crop transplanter shown in Fig. 6.

Fig. 6: Photograph of a tomato plant exiting the transplanter with its accompanying plant label that has been coated with the fluorescent orange crop signalling compound.

Weed control mechanisms

While there are many approaches to weed control, most weed control machines typically employ either mechanical or chemical weed control methods. These approaches have been used in conventional mechanized agriculture for many years, and recently, they have been coupled with automation technology to either reduce inputs or exert more precise control of weed plants. While other weed control mechanisms exist, such as flaming, hot water or steam, or high voltage, the adoption of these technologies has been low, and little research has been reported to automate them. Thus, this section will focus only on chemical and mechanical weed control mechanisms.

Selective chemical application

Selective spraying systems, sometimes called spot spraying systems, turn nozzles on or off based on what is at nozzle locations. In the case of weed control, once a selective chemical application system perceives the existence of weed plants, herbicide can be applied to the area surrounding the weed locations and not applied in areas where no weed plants exist. This selective chemical application strategy selectively switches nozzles on and off based on presence or absence of weeds and has been investigated for some time in the context of conventional spray application systems. There are some limited examples of commercialized products on conventional spraying platforms such as the WeedSeeker, which uses photodetectors and an active light source to detect vegetation between crop rows. Machine vision-based selective patch spraying has been investigated for row crops that could deliver herbicide onto areas where weeds are present while travelling at typical operational speeds. More recently, selective spraying system are emerging, often in the context of agricultural robotics.

There are also examples of work to increase the spatial resolution of the nozzle with the idea that the spray pattern is intended to be directed to individual plants. Researchers, for example, developed and tested a prototype pulsed-jet, micro-dosing actuator capable of applying microliter herbicide dose rates to small target areas as small as 9 mm x 12.7 mm. This small target area was achieved with eight micro-spray nozzles each consisting of five hypodermics each forming a micro-spray nozzle array. Eight micro-spray nozzles were spaced horizontally at a 1.27-cm interval and were located at a 10.16-cm nozzle height. Similarly, Researchersreported a system that applied heated oil to weeds using a micro-spray manifold in 6.3 mm by 12.5 mm rectangular areas. This high-spatial-resolution approach to selective chemical application should substantially reduce the volume of applied chemicals to the field over conventional low-resolution nozzles. It may also enable the use of nonselective herbicides to control weeds in the context of high-value vegetable crops.

Other examples of selective spraying include a selective sprayer which sensed tall weeds in blueberries using an ultrasonic sensor and selectively applied herbicide to the weeds. Researchers described the development of a selective sprayer designed to control mesquite in road right-of-ways. Infrared light was transmitted horizontally to a receiver. When this beam was interrupted, the sprayer sensed the presence of a mesquite plant and correspondingly applied herbicide to the area nearby and containing the plant. Both of these examples demonstrate the concept of using basic sensing technology to perceive a highly distinguishable feature of weed plants to determine their presence. Additionally, there are examples of selective spraying for other pests.

Robotic selective application

While selective herbicide application systems show tremendous promise in reducing the volume of chemicals applied, they by themselves should probably not be considered to be robotic, rather they are automated control systems that respond to perceived weeds in the field. They generally have not included much machine behavior layer technology to deal with uncertain situations. However, more recently, selective spraying technology has been coupled with smaller, automatically guided vehicle technology with hardware and control technology to more precisely apply herbicide. For example, the ecoRobotix spraying robot is autonomously guided through crop rows using GPS and machine vision sensors to follow crop rows. This robot detects weed plants, and then uses two spray nozzles on delta robot arms to position two nozzles over weed plants and selectively apply herbicide directly to detected weed plants (Fig. 7).

Fig. 7. The ecoRobotix robot senses weeds in the field and then moves spray nozzle to the weed to achieve goal of directly delivering herbicide to each weed plant.

The Australian Centre for Field Robotics' Ladybird robot is similar to therobot, but used a six-axis robot arm (model UR5, Universal Robots, Odense, Denmark) to move a spray nozzle end effector to the weed plants (Fig. 8).

Fig. 8 Ladybird robot features a novel design covered with solar panels (left) and a selective spraying system in which the spray nozzle is moved close to weed plants by a six-axis robotic arm (right).

Mechanical weeders

For situations where chemical weed control is not consistent with the producer's management practices, mechanical cultivation or tillage is often used. There is a wide variety of mechanical tool designs for mechanical cultivation that can be used with both interrow and intra-row weeds. Generally, they rely on three main physical techniques for controlling weeds which include burying, cutting, and uprooting the weed plants. Each of these techniques interferes with the growth of weed plants by killing them or slowing their growth so that the crop plants can overtake the weed plants achieving greater canopy closure and reducing the light interception by the weed plants. Mechanical cultivation is used widely in organic production systems and has been investigated with automated control.

A review of the research and commercial literature found that the majority (75%) of robotic weeders used mechanical over chemical weed control. Of the mechanical types of solutions, there were generally two main classes of approaches: (1) passive cultivation tools being automatically guided through crop rows, and (2) active control of weeding tools in the row and sometimes between the row.

Examples of automatically guided cultivation tools are Naïo Technologies Oz weeder and Dino robot (Naïo Technologies, 2019a,b). The Oz weeder is electrically powered with four electric motors, one in each wheel. It comes in a small package that is 60 cm high and 40 cm wide, weighing 110 kg. Because of its size, it fits between rows of vegetable crops and carries a variety of cultivation tools including several types of harrows or brush or spring weeders for intra-row weeding. The Dino weeding robot is a larger platform with a 130 cm height and adjustable wheelbase from 140 to 180 cm. It is able to span 120 cm of row width, and guides different passive cultivation tools through multiple rows. The Kongskilde Robotti has a similar platform spanning multiple rows and is automatically guided to cultivate the crop rows that it spans.

The systems that apply active control of mechanical tillage tools are typically using local scene information from the perception system to actively move the tools into and out of the crop row depending on where crop plants are located. Following this basic strategy, different systems have taken different approaches largely based on their mechanism design.

One common approach is to use a horizontal knife or blade hoe tool that is positioned shallowly below the soil. The blade is moved into the crop row in the spaces between the crop plants cutting and burying weed plants. An early implementation of this strategy was the Sarl Radis weeder developed in France. This automated weeder detected reflected light from the crop plants, and accordingly moved the hoes out of the row when a crop plant was detected. Originally developed for transplanted crops, it had highest performance when the weeds were substantially smaller than the crop plants.

There are examples of novel intra-row weeding tools. A cycloid hoe concept was developed and reported . The cycloid hoe consisted of eight tines that rotated around a vertical axis. The hoe mechanism was designed so that individual tines could go into the row in the absence of a crop plant or be retracted when a crop plant was present.

The system was tested with artificial plants and georeferenced crop plants and was found to be reliable in performing intra-row weeding without interfering with the crop plants. As an example of another mechanism, researchers described the development of a weed stamping tool. To injure weed plants, the tool is inserted 47 mm into the soil while centered on individual weed plants. The stamping tool was mounted on a delta robot arm that positioned the tool over detected weed plants. In field tests, 85% of the plants that were stamped died. Researchers investigated the effectiveness of a rotating tine mechanism in disturbing simulated weed plants with tine depth and rotation speed as experimental factors.

WEED AND CROP CLASSIFICATION ALGORITHM

An automatic, robust, real-time, machine vision-based system for intra-row weed knife path control was created to detect and locate the crop signalling compound on the plant labels. The system was designed to work outdoors in a vegetable field environment with heavy weed density, and to identify occluded and non-occluded crop plants of assorted sizes by their labels. Figure 9 shows an example weedy field in one of the experiments with lettuce plants and their plant labels where the lettuce was occluded by different kinds of weeds of various heights, colours, and shapes (Fig. 9b). All the rows were planted in raised beds at a tractor travel speed in the range of 1.6 to km h-1.

A real-time, 3D geometric appearance-based algorithm for detecting and locating the crop signalling system was developed. A fluorescent image of the field was electronically captured in real-time using the camera under UV illumination to find the crop signal. Camera exposure time was set to 4 ms to have sufficient image intensity from the 10-Watt UV LEDs while also eliminating motion blur from the moving sled. In each image, six separate regions of interest (ROIs) were defined based on the locations of the six mirrors in the image. Because of the very heavy weed densities, the level of visual occlusion was severe and the crop signal, in most cases, could not be detected correctly using a single top view, due to its occlusion by foliage. Additionally, it was not possible to observe the bottom most point where the crop plant entered the soil. By utilising the six side-mirrors, however, it was possible to detect the crop signal and estimate the point where the plant label entered the soil.

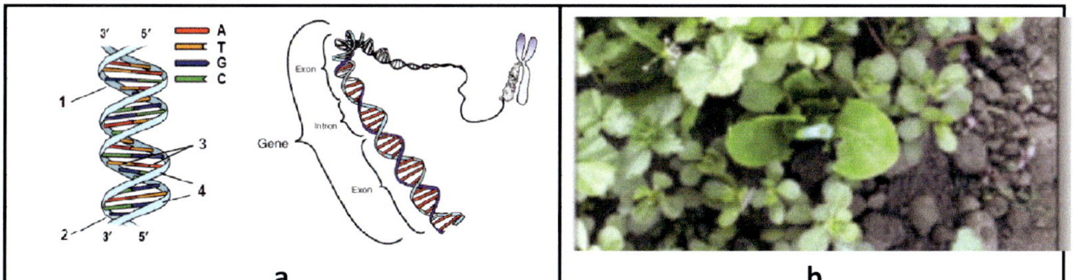

Fig. 9: (a) A weedy field of lettuce plants and their green plant labels, where (b) the lettuce was occluded by different kinds of weeds of various heights, colours, and shapes.

METHODS OF OPERATION

Robotic weeders come in a range of shapes and sizes, but in general they can be divided into those which deliver chemical herbicides and those which automate traditional processes of mechanical weeding such as hoeing.

The first type of 'robots' are those which are still manually operated, and designed to reduce the amount of labour required and change the mode of that labour from the physical task of removing weeds by hand, to the more efficient method of directing a machine to do so. They can be thought of in the same manner as tractors, or combine harvesters – as labour saving devices. Many are self-guiding using simple odometry sensors or GPS technology, requiring only adjustment to the desired route, or factors such as how much herbicide to apply, or how vigorously a mechanical hoe operates. Such devices may not deserve the title of robot, since they do not act without human intervention, a characteristic which commonly defines a 'true' robot.

Weeders such as Lettuce Bot (Blue River Technology), now well established in the US, are pulled over the top of a vegetable crop by a tractor, and utilises computer vision equipment to compare an image of every plant encountered with an extensive database of control images. Upon detecting an unwanted plant, the machine delivers a precise application of concentrated fertiliser (or herbicide if desired), killing the weed but enriching the nutrient content of the soil for the surrounding crop.

There are equivalent mechanical systems such as Robovator (F. Poulsen Engineering), which use the same computer vision technology to avoid the crop, whilst deploying tools to disturb the soil. In such machines, planting plans can be pre-loaded and the tool used to deliver the weeding action can vary from a mechanical hoe to a thermal weeding tool utilising a pinpoint flame or heated oil, a targeted laser device or a combination of tools for different areas of soil.

These robots are considered to reduce the labour requirement for weeding by a factor of 10. They operate at 2-4 mph and may cost somewhere between £100,000 and £200,000 to buy, depending on their specifications. Their computer vision tools can be combined with seed/seedling mapping, and allow them to remove weeds closer to the crop.

A new range of bots, such as Ecorobotix (Ecorobotix Ltd.), take similar tools but remove the need for manual control, instead adding a wider range of sensors, and automating the process using complementary technologies such as solar power. Much like solar-powered lawn mowers, these machines are able to be programmed with a route to follow or guide by remote technologies, given a schedule to operate upon, and then left to complete the task unaided.

Technologies

Evaluating the potential to use a robotic weeder in any given production system, requires a robust understanding of the technology it utilises, its potential benefits, and its limitations.

A range of technological solutions to automation in agriculture are tested as part of the Hands-Free Hectare, an innovative test case for future production methods.

Where it goes

The first level of technology is that which decides where the robot can go. For labour saving devices, that may be as simple as following the tractor to which it is attached. However, at the next level of automation, robots acting independently can be guided by real time kinematic global position systems (RTK-GPS), which use satellite navigation techniques (an advance on a car's 'Sat. Nav.'). There are examples of weeders using these systems such as DINO (Naio Technologies).

The system allows the robot to act without supervision, assuming that an accurate map of the area is available. These systems can be coupled with seed/seedling geo-referencing, which allows the robot to be guided by the position in which the seeds were originally sown. This relies on precision sowing and the ability to store and retain accurate records.

What it 'knows'

The ability to refine the application of robotic weeding depends on how much information is available upon which to base decisions. Robots commonly contain a multitude of remote sensors, collecting information on environmental factors such as meteorology, soil parameters, and on-board activities.

The accuracy of this information, its interpretation, and the ability to record it, each contribute to informed decision making, either by a person controlling the use of a robotic weeder, or in the future by the robot itself, either responding to changes or applying situation specific settings. The sensors can also be used to monitor general information about the production operation which, even if they are not used to directly adjust the performance of the robot, can still be informative at a farm business level.

HOW IT 'SEES'

The robot needs to be able to differentiate between a weed and a crop plant. This is the basic difference between a simple tractor mounted between-row weeder, and a precision robot capable of weeding within rows. Plant identification relies on the ability to assess biological morphology, spectral characteristics (colour/reflectance) and texture. The use of biological morphology relies on shape recognition, and the fact that each species has a unique shape, something which is particularly challenging for crops which are closely related to weed species. Advances on these technologies include a transition from 2D to 3D cameras, multispectral imaging and plant identification libraries containing many thousands of images of crops and weeds from every conceivable angle to act as a reference point.

How it acts

Once a weed is detected, the robot must take action. The way it does this has an impact upon its capacity and accuracy. If it is applying a chemical treatment, it must be able to do so selectively and accurately using advanced micro-spraying technologies and highly accurate spray nozzles. This not only ensures satisfactory weed removal but also contributes to the environmental benefit of using a robot. If the action delivered is mechanical in nature then the type of tool, and its sensitivity are the critical factors. A range of precision tools are utilised on robotic weeders including hoes, finger tines, oscillating discs and cutters. There are also examples of other precision tools applying flames, freezing, lasers, air blasts and even 160°C heated food grade oil to kill weeds.

What it needs

In order to operate, a robot needs a power source, and a maintenance regime. Robotic weeders can be operated by being towed, but there are also examples with diesel engines, battery power (chargeable) and on-board energy generation through solar panels. The source of power may determine somewhat, the ability of the robot to act autonomously, but also the need for a fuel source and the associated costs. The ability of the robot to detect problems such as blockages, and fix them will also impact upon autonomy, whilst the requirement for maintenance may result in labour costs which are removed from the field, being reapplied in the farm yard.

If it learns

The critical ability, which could lead to genuine labour free weeding, is the ability of a robot to not only measure and record information, but to act upon it, and potentially, to actively learn from experiences to improve performance. At present, this factor is reduced to the 'intelligence' level of the robot, i.e. its ability to make decisions based on the information it receives rather than to rely on the direction of a human supervisor. This could include elements such as an ability to detect its own inefficiency and repeat an action, rather than relying on follow-up manual labour, or to pause activity during a heavy rain shower.

DRONE AND SENSOR TECHNOLOGY FOR SUSTAINABLE WEED MANAGEMENT

Biotic threats such as insects, weeds, fungi, viruses, and bacteria can broadly affect crop yield and quality. Among these, weeds are the most impacting problem causing remarkable yield loss worldwide. The most characterized effect of weeds is competition for resources such as light, water, space, and nutrients. In addition, specific chemical signals and/or toxic molecules produced by weeds may interfere with a normal crop development. A distinctive trait of wild species, including weeds, is their high physiological, morphological, and anatomical plasticity which makes them more tolerant than crop species to environmental stressors.

Moreover, weeds interact with other biological components of the environment, acting as refuge for plant pests such as insects, fungi, and bacteria that can harm close in crops. For example, wild oats (Avena fatua L.) can harbor the etiological agents of the powdery mildew in crops such as wheat (Triticum aestivum L.), oats, and barley

(Hordeum vulgare L.); altamisa (Parthenium hysterophorus L.) can be a secondary host of the common hairy caterpillar (Diacrisia obliqua Walk.); Cyperus rotundus can host the root-knot Meloidogyne graminicola and, therefore, can contribute to their spreading in the field.

Finally, weed infestation may affect fresh and processed products quality such as beer, wine, forage. In this respect, weed residuals may cause accumulation of off-flavors products, or in some cases, can make them harmful to humans and animals. Weeds may also contain high levels of allergens and/or toxic metabolites that, if ingested, can cause asthma, skin rash, and other reactions. Most weed research aims at developing strategies that can reduce the deleterious impact of the interspecific competition between crops and weeds and recent technological advances may further contribute to this scope, while improving the sustainability of weed control.

Worldwide, weed competition causes severe yield reduction in all major crops, such as wheat (23%), soybean (37%), rice (37%), maize (40%), cotton (36%), and potato (30%). Yearly, weeds cause 50% yield losses of corn and soybean productivity in North America. For corn, this equates to a loss of 148 million tons for an economical loss of over $26.7 billion.

In Australia, yield loss due to weeds accounts for 2.76 million tons of grain from different plants, including wheat, barley, oats, canola, sorghum, and pulses. The annual global economic loss caused by weeds has been estimated to be more than $100 billion U.S. dollars, despite worldwide annual herbicide sales in the range of $25 billion. In Europe, herbicides are the second most-sold pesticides. They accounted for 35% of all pesticide sales in 2018, overcoming insecticides and acaricides

Weed management requires an integrated approach

In 2050, the world population will quadruplicate, reaching 9.15 billion people. However, the predicted increase in food demand will be hardly met by the current production system. Also, climate change will be an additional challenge for the human food supply in the near future. Among all the processes affecting crop productivity, weed management will be one of the hardest challenges. Mechanical and chemical weed control has disadvantages that probably will impede them to be effective for future weed management. Mechanical methods are scarcely efficient, and herbicides have a high ecological impact. An approach that minimizes the drawbacks of mechanical and chemical weed control is Integrated Weed Management (IWM). IWM combines chemical, biological, mechanical, and/or crop management methods, and represents a model to improve the efficiency and sustainability of weed control. In contrast to traditional methods, IWM integrates several agro-ecological aspects such as the role of conservation tillage and crop rotation on weeds seed bank dynamics, the ability to forecast the critical period of weed interference and their competition with crops, and the specific critical levels of crops/weeds interaction. Therefore, an effective IWM must rely on a thorough knowledge of crop-weeds competition dynamics, which currently represents one of the most active research areas in weed science.

New technologies for site-specific weed management

Precision agriculture relies on technologies that combine sensors, information systems, and informed management to optimize crop productivity and to reduce the environmental impact. Nowadays, precision agriculture has a broad range of applications and it is employed in different agricultural contexts including pests control, fertilization, irrigation, sowing and harvesting. Precision agriculture can be effectively applied to IWM also.

In the last decade, precision agriculture has rapidly advanced because of technological innovations in the areas of sensors, computer hardware, nanotechnology, unmanned vehicles systems and robots that may allow for specific identification of weeds that are present in the field. Unmanned aerial vehicles (UAV) are one of the most successful technologies applied in precision agriculture. Unmanned Vehicles systems are mobile Aerial (UAV) or Terrestrial (UTV) platforms that provide numerous advantages for the execution and monitoring of farming activities. UAVs can be highly valuable since they allow for Site-Specific Weed Management (SSWM) (Fig. 10).

SSWM is an improved weed management approach for highly efficient and environmentally safe control of weed populations, enabling precise and continuous monitoring and mapping of weed infestation. SSWM consents to optimize weed treatments for each specific agronomical situation. The combination of UAVs with advanced cameras and sensors, able to discern specific weeds, and GPS technologies, that provide geographical information for field mapping, can help in precisely monitoring large areas in a few minutes. Thanks to more accurate planning of weed management that can increase mechanical methods effectiveness and/or reduce herbicide spread, the potential agro-ecological and economic implications of SSWM are remarkable, yielding lower production costs, reducing the onset of weed resistance, improving biodiversity, and containing environmental impacts. The application of UAVs to weed control can, therefore, contribute to improve the sustainability of future agricultural production systems that must comply with an increasing world population.

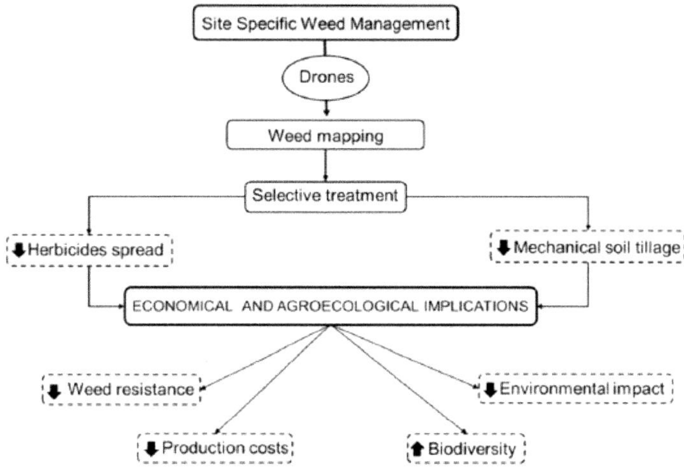

Fig. 10 Site-specific weed management (SSWM) scheme realized by drones and its economical and agro-ecological implications

UAVs remote sensing techniques and sensors

UAVs have become a common tool in precision agriculture. Thanks to their affordability, user-friendliness and versatility, UAVs are often the primary choice for fast and precise in situ remote sensing or survey operations. Despite their versatility, these systems may be used for different purposes, depending on the sensors they carry on. Ongoing research is looking at the best solutions to integrate data collected from sensors on UAVs, ground sensors and other data sources for better management of punctual operations in the field, with a particular focus on smart agriculture and big data management.

Although UAVs systems do not offer the same territorial coverage as satellites, they offer a spatial and temporal resolution that other systems do not. From an economic point of view, the use of drones requires the investment to buy a UAV system with at least a 0.1 cm/px resolution RGB camera, a trained pilot for flight management and post-processing software capabilities. The initial UAV investment is compensated by the repeatability of flights, which increases the frequency of datasets delivered, and the higher resolution compared to other systems. UAVs systems also have further advantages: (1) the possibility to collect easily deployable data in real time (excluding post-processing); (2) they can be used to survey areas with high level of hazard and/or difficult to reach; (3) they allow operators to collect data even with unfavorable weather conditions, such as in very cloudy or foggy days, under which satellite detection systems fail or produce very altered datasets. The most important sensors available as payload are mainly categorized into three classes depending on the spectral length and number they can record:

- RGB (Red, Green, Blue) or VIS (Visible) sensors
- Multispectral sensors
- Hyperspectral sensors

RGB/VIS sensors

The RGB or VIS sensors are the most common and largely available commercial cameras (Table 1). Their possible applications have been the focus of most research for years due to their potential and low-cost operational requirements

These sensors are used to calculate vegetation indices such as the Green/Red Vegetation Index (GRVI), Greenness Index (GI) and Excessive Greenness (ExG) with acceptable or high levels of accuracy. Also, RGB sensors have been increasingly used for machine learning techniques in object recognition, phenology, pathologies, and similar purposes. The typical workflow of processing RGB images from UAVs for remote sensing is: 1. pre-flight planning, 2. flight and image acquisition, 3. post-processing and indexes or dataset extrapolation.

Phase 1 is critical and essential to collect data of useful quality for the purpose. In the pre-flight planning phase, the parameters to consider are the definition of the study area, the flight altitude, site topography, weather forecast and local regulations for unmanned flights.

In phase 2, it is recommended to keep the data flow sufficient to store data and to check if the acquisition platform can acquire the amount of data required. It could be possible to encounter I/O errors due to the inadequacy of the platform with consequent loss of information or abortion of the mission.

In phase 3, for RGB sensors, there is no need to perform radiometric calibration, which is the case when using multispectral and hyperspectral sensors. RGB data can be used per se or to create a georeferenced orthomosaic. In this case, the individual images are rectified, georeferenced using GPS data and stitched together to form a single image (orthomosaic) covering the entire study area. Orthomosaics can be generated either with RGB values as they are or after calculating the desired vegetation indices. If RGB images are to be used in machine learning algorithms, the workflow is different. In this case, it is necessary to collect a large dataset of images for the training and testing of the algorithm. This dataset may already be available from third-party sources, such as PlantVillage or PlantDoc. Alternatively, it can be created from scratch if the purpose of the research is not covered by existing datasets. In this case, the acquisition, selection and processing of the images are critical, because the final dataset can affect both the training and the use of the neural network, with risks of producing biased results.

Table 1 RGB cameras and their main specifications

Camera model	Sensor type and resolution [Mpx]	Sensor format	Sensor Size mm]	Weight [kg]	Price (approx [€]
Canon EOS 5d Mark IV	CMOS 30.4	Full Frame	36.0 × 24.0	ca. 1.0	ca. 1000
Nikon D610	CMOS 24.3	Full Frame	36.0 × 24.0	ca. 1.250	ca. 1000

Sony Alpha 7R II	CMOS 42	Full Frame Mirrorless	35.0 × 24.0	ca. 0.6	ca. 1200
Sony Alpha a6300	CMOS 24	Small Frame Mirrorless	23.5 × 15.6	ca. 0.8	ca. 800
Panasonic Lumix DMC GX8	CMOS 20	Small Frame Mirrorless	17.3 × 13	ca. 0.5	ca. 1000
Panasonic Lumix DMC GX80	DLMOS 16	Small Frame Mirrorless	17.3 × 13	ca. 0.5	ca. 500
DJI Phantom 4 Pro *	CMOS 20	Small Frame	13.2 × 8.8	ca. 1.5 (with UAV)	ca. 1500 (with UAV)
DJI Mavic 2 Proa	CMOS 20	Small Frame	13.2 × 8.8	a. 1.5 (with UAV)	ca. 1500 (with UAV)

Multispectral sensors

The multispectral sensors are used for a wider range of calculations of vegetation indices as they can rely on a higher number of radiometric bands. A comparison of the most common multispectral sensors, specific for UAV systems.

With multispectral sensors, the range of vegetation indices that can be monitored is considerably extended compared to those that can be calculated with only three RGB bands. Moreover, the workflow has minor variations. For these sensors, in phase 1, the radiometric calibration and atmospheric correction phases are strictly required. Many multispectral sensors, such as the Micasense RedEdge series or the Parrot Sequoia +, have downwelling irradiance sensors and a calibrated reflectance panel to address some of the requirements for radiometric calibration. Due to a lower resolution of the sensors compared to RGB ones, a lower flight altitude and an adequate horizontal and vertical overlap of recorded images must be taken into account to obtain an adequate ground resolution for the surveyed objective and to avoid missing data.

In phase 2, having a higher number of radiometric bands to record, the dataflow will be higher so is critical to avoid I/O errors, missing data or mission failures. Due to multi-lenses nature of the sensors in phase 3, the data collected suffer from the parallax problem. As a consequence, images have to be rectified, georeferenced and must be stacked to generate a single image with different radiometric levels, and calibrated with the downwelling irradiance sensors data acquired during the flight. After this procedure, it is possible to generate a multispectral orthomosaic and then calculate the requested indexes. Multispectral images are also used in machine learning applications taking into account the multi-camera nature of sensors and the different bands recorded. Thanks to the availability of a higher number of radiometric bands, the machine learning algorithms can be extended to not-visible recognition such as early stage plant disease, field quality assessment, soil water content, and more.

Hyperspectral sensors

The hyperspectral sensors can record hundreds to thousands of narrow radiometric bands, usually in visible and infrared ranges. To deal with hyperspectral applications, the choice of number and radiometric range of bands is critical. Each band or combination of bands, being very narrow, can detect a specific field characteristic.

Each hyperspectral sensor can detect only a certain number of bands, so the aim of survey must be very clear to choose the right sensor. Although hyperspectral sensors have decreased in price in recent years, they are still an important starting investment since they are much more expensive than RGB and multispectral sensors. In addition, they are heavier and bigger than other sensors, often making their use on UAV systems difficult and/or excessively onerous in terms of payload. Some of most used hyperspectral sensors in UAVs application and their main characteristics are shown in Table 2.

Applications of UAVs to weed management

UAVs are ideal to identify weed patches. The main advantages of UAVs compared to UTVs are the shorter monitoring/surveying time they require and optimal control in the presence of obstacles, which is critical when working between crop rows. In a few minutes, UAVs can cover many hectares flying over the field, thus providing the photographic material for weed patches identification. These images are processed via deep neural network, convolutional neural network, and object-based image analysis. Based on a systematic review of the literature concerning weed identification by UAVs, it can be concluded that mainly three types of cameras are used for weed patches identification: RGB, multispectral and hyperspectral cameras.

These cameras are very similar in terms of information obtained for the purpose of weeds identification. Indeed, the three camera types can recognize weed patches with good accuracy depending on flying altitude, camera resolution and UAV used. UAVs have been mainly tested on important crops such as Triticum spp., Hordeum vulgare, Beta vulgaris, Zea mays. These are among the most cultivated crops worldwide and are highly susceptible to weed competition especially in early phenological stages. In these crops, it was possible to identify several dicotyledonous weeds including Amaranthus palmeri, Chenopodium album and Cirsium arvense, as well as different monocotyledonous such as Phalaris spp., Avena spp. and Lolium spp.. These weed species are widespread globally and can be a serious threat to different crops [107, 108]. Therefore, the combined use of UAVs and image processing technologies may contribute to effectively control different weed species interfering with the crops with relevant environmental benefits.

PRACTICAL CONSIDERATIONS

If only the task at hand, namely weeding a row of vegetables is considered (without any social bias) then the main disadvantage of a robot is its inability to be reactive to complex and highly variable environments that are typical of real-life production systems. Whereas a human can alter their activity to suit the imprecise nature of fields which are not square, flat, or necessarily precision planted, a robot cannot, at present, make such adjustments. This limits their application to row crops, which are precisely planted, and prevents their effective use on broadacre crops, closed canopies or on land which undulates. The limitations of using such a system, in a Welsh small-scale horticultural context are currently being investigated.

Camera model	**Lens**	**Spectral range [μm]**	**Spectral bands [number and μm]**	**Weight kg]**	**Price (approx.) [€]**
CUBERT	Snapshot+PAN	450–995	125 (8 μm)	ca. 0.5	ca. 50000
Cornirg microHSI 410 SHARK	CCD/CMOS	400–1000	300 (2 μm)	ca. 0.7	–
Rikola Ltd. hyperspectral camera	CMOS	500–900	40 (10 μm)	ca. 0.6	ca. 40000
Specim-AISA KESTREL16	Push-broom	600–1640	350 (3 – 8 μm	ca. 2.5	–
Headwall PhotonicsMicro-	InGaAs	900–1700	62 (12.9 μm)	ca. 1.1	–
hyperspec X-series NIR					

There are limits too, to the use of certain technologies. Computer vision works well on a clean, green leaf, but it cannot make an accurate determination if the leaf is dirty for example. The accuracy of differentiation between weed and crop is also technically challenging, particularly at the seedling stage, which means a loss is expected for mistakes, or inadvertent damage caused by imprecise application of mechanical tools. In these scenarios there are fine margins of error, particularly at relatively high speeds. Where there is even spacing and predictability, the robot can cope, but any deviation from that creates the possibility of accidental crop damage, or the need to repeat the process by hand.

If a robot applies a chemical treatment, it too is limited by the availability and regulation of effective chemicals. Whilst it may offer environmental benefits through the precise application of a given chemical, it could still fall foul of changes in regulations or policy which prevent their use. Furthermore, specific crops may have a limited range of approved treatments (e.g. spinach) which may limit the ability of a producer to use the same robot for multiple crop types.

Producers must also evaluate the economic case for using a robotic system. Whilst the pressures outlined in this article apply universally, there is still no guarantee that a robot would prove more cost effective than hand weeding, or that the need for labour would be eliminated. A comparative study of labour requirements between hand hoeing and robotic hoeing reported that the average number of person hours needed to weed 100 m2 was 0.241. Once the robot had passed, the number of 'follow-up' person hours needed to render the plot completely weed free was 0.102, a reduction of 57.5%. This highlights the fact that a robot does not necessarily replace labour, but may simply reduce it, and although this figure would improve with each pass of the robot over the same row, it may not be able to achieve the same efficiency as a person. The cost of the equipment must also be factored in, for example the previously featured 'Robovator' could cost £100,000 to buy and around £213 per Ha to operate. Whilst the operation cost may be roughly a tenth of the cost of hand weeding the same area, the costs associated with purchase, maintenance and depreciation are additive, and unique to the robot.

The factors which are unique to the specific farm business are perhaps the most important considerations. Robots cope best with well-established stands, low weed densities, and crops which are larger than the weeds. Any deviation from this ideal scenario is likely to make operations less efficient, and either require more equipment (and therefore cost) or more follow-up labour. Robots are also, most effective before weeds become established, and early interventions are considered crucial for a high degree of success. Meeting these specifications may mean that, for the time being, robots are best suited to speciality crops, such as asparagus or lettuce. Research has also highlighted the importance of intelligence in the robotic system (the ability to accurately detect the weed and/or the crop), essentially this means that the more equipment the robot has, the closer it can get to the crop, and the more effective it can be. This indicates a diminishing return with simpler, less costly technologies. The operation and programming of robotic systems requires fewer human hours, but the ones it does need require a high degree of technical competence, skills which are currently in short supply.

FUTURE OPPORTUNITIES

The future of robotic agriculture rests upon the ability to produce 'true' robots, machines that are entirely independent of human control. This transition from processes which are more automated to processes that are fully automated will typify the evolution from what is referred to in the literature as agriculture 4.0 to agriculture 5.0. The key technological advancement that would enable this to become a reality is the development of genuine artificial intelligence – the ability for a robot to learn. In a recent publication, a weeding robot was considered truly robotic if it met the following criteria:

- It had the ability to monitor the crop, weeds, weather and soil in real time.
- It could decide when the crop needed to be weeded.
- It could choose the optimal weeding implement.
- It could take the weeder to the field that needed to be weeded.
- It could adjust the chosen implement for optimum performance.

- It could monitor itself for blockages and mechanical problems, and fix them.
- It could continually monitor and adjust performance in the field.
- It could return the weeder to the farm once the task was complete.
- It could clean, maintain and store the weeder.

This list simply seeks to highlight the difference between a labour saving device, and something truly robotic, and that there is a large difference between them. However, most of the technology required to achieve this already exists, particularly in terms of remote sensing. With the ability to 'learn' and apply that learning to variable contexts, robotic systems may well be able to replace labour to a much higher degree than is currently possible.

Near market improvements in robotic weeding include improved weed/crop differentiation through rapid advances in image analysis techniques and the potential for crop tagging. Alongside this are improvements in precision instruments, able to get closer to the crop, with minimal disturbance and damage. Robots also have the potential to be quickly updated though software, which can make them more responsive to changes in technology and trends for data driven decision making brought about by the advent of the internet of things.

Skills training, policy making and regulatory mechanisms need to keep pace with the capabilities of technology for them to be efficiently utilised. The most commonly quoted milestone for the predication of human activity, particularly in terms of agricultural output is 2050. Any consideration of weed management in 2050 will need to be able to take the early promise of today's robotic labour saving devices and apply them to the most complex production systems as well as the most extensive. It has recently been reported that an entirely different approach would be required for broadacre crops, and to achieve truly sustainable intensification, that is the field on which the robots must next do battle.

11

AUTOMATED CROP DISEASE MONITORING AND MANAGEMENT

The agriculture sector plays a noteworthy role in the development of the country's economy as such, it provides large-scale employment opportunities to people in countries like India, Colombia, USA, China, Brazil, etc. So, the growth of the agricultural sector is necessary for the development of the economic condition of a country. The UN Food and Agriculture Organization has reported that the world has to produce 70 percent more food in 2050 than it produced in 2006 to meet out the food demands of the growing population across the world. With the advent of the advanced technologies in both the agricultural industry and, information and communication technology, human society can produce enough food to meet the demand of more than 7 billion people.

However, there exists a threaten to food security by a number of factors including climate change, the decline in pollinators, plant diseases, and some other factors. Plant diseases are not only a threat to food security at the global level but also cause disastrous for smallholder farmers whose livelihoods rely on healthy crops. In the developing countries like India, more than 80 percent of the agricultural production is generated by smallholder farmers, and a report suggests that there is a loss of more than 50% in yield due to pests and diseases.

Moreover, a study reveals that around 50% of the people live in smallholder farming households making smallholder farmers as a group, that is, particularly vulnerable to pathogen-derived disruptions in food supply. So, this is the right time to improve productivity and enhance the quality of the agricultural products, unless otherwise, the people will face food scarcity in the future across the world. But, unfortunately, the farmers face a lot of difficulties in day-to-day crop monitoring and management system like more use of pesticide and fertilizer, disturbance of insects, some types of plant diseases, food preservation, rainfall water scarcity, and so on. Particularly, nowadays, varieties of diseases attack crops so that it creates a huge amount of loss in yield and the farmers have to spend more money on curing the attack. Moreover, many farmers still use the traditional method of farming, which results in low yield; it could also be one of the main factors for the low yield.

AGRIROBOT

Here we proposed aagri-Robo which capture the leaf images in real time and detect and classify infection in particular field. For this, we use MATLAB software to run the disease detection algorithms on pc.\

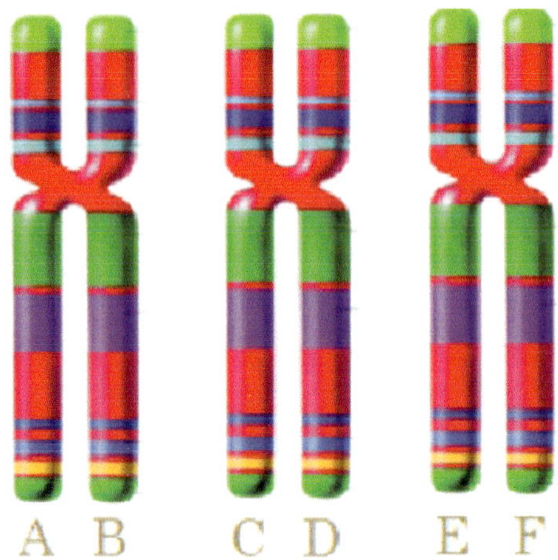

Fig.1. Agri-robo

The Robot also have injection mechanism, hence it is completely autonomous robot. These systems consist of camera placed on robotic car. This robot is made in such a way that it can move around the farm in space, in space between two row of crops and camera take the images of plant. The camera is connected to PC wirelessly on which main process of disease detection is run.. PC is connected to robot through RF module for wireless control of agrirobo from certain distance. After the detection of disease a spraying system which is designed on robot automatically spray the pesticides in proper amount and where it required.

ALGORITHM FOR DISEASE DETECTION USING IMAGE PROCESSING

An efficient and speedy image processing algorithms was developed using MATLAB software. The first phase of algorithms deals with separation of healthy crop from affected one latter phase of algorithms concentrates on finding the disease on crop.

The steps required in image processing system:

- Load the images
- Enhancement
- Transfer the input image from RGB to LAB pattern

- Masking of green-pixels
- Removal of masked green pixels
- Image segmentation
- feature extraction

In image processing, sometimes we have to convert the image from one colour space to other. MATLAB contain tools require for image processing which covers all colour space transformation.

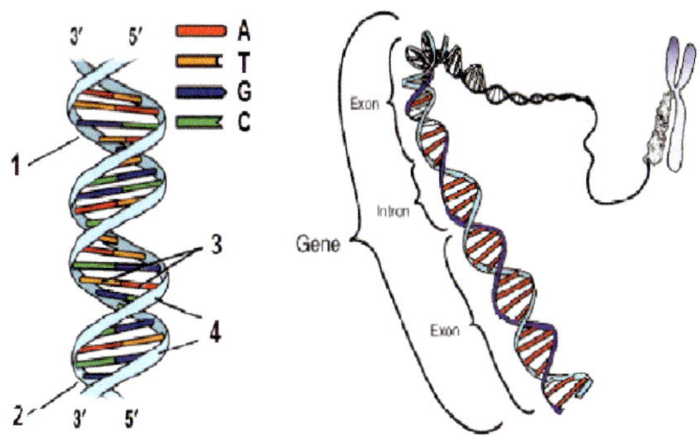

Fig 2 Flowchart of algorithm steps following in image processing

Input color images have primary colors green, blue and red. It is not possible to implement the applications using RGB because of their range i.e. 0 to 255. Hence they convert the RGB images into the LAB images. LAB consist of L*, a* and b* layers. L* represent brightness represent degree of redness- greenish. B*represent degree of yellowish bluish.

Convert image to L*a*b* color space

Cform 2lab = makecform(srgb2lab);

LAB = applycform(RGB, cform2lab);

In masking process green pixel are find out and given out the zero pixel value. Some threshold value is set for detection of disease. The pixel having value less than threshold are set to zero and if it has high value then set to one. In segmentation part of leave having same features are divided into number of segment. The diseased part is then segmented into a number of small segments of same size. segments are obtained by segmentation process. The size of the segment must be selected in such a way that the proper information is not lost. From the above steps, the diseased part of the crop is find out. Not all segments consist useful quantity of information. Hence the patches which contain more than fifty percent of the useful information are taken as for the further analysis. If the image is in RGB form converts it into grey colour form with the help of following equation,

G(x) =0.2989*R + 0.5870*G + 0.114.*B

After segmentation the infected region various features are extracted to describe the infected region. Colour, texture and shape based features are normally used for region description. Final stage in disease detection is classification. It identifies a rule according to selected features and assigning each disease to any one the predetermined classes. Once the feature extraction was complete, two files were obtained.

They were: (i) Training texture feature data, and (ii) Test texture feature data. A software routine was written in MATLAB that would take in .mat files representing the training and test data, train the classifier using the train files, and then use the test file to perform the classification task on the test data. Consequently, a MATLAB routine would load all the data files (training and test data files) and make modifications to the data according to the proposed model chosen. These both file are compared with each other and corresponding disease is detected. How we can detect the disease is explained by the above flowchart.

SPRAYING TECHNIQUES

In our system the captured images from camera processed in PC using image processing technique and the detection of disease is done and corresponding signal is given to the microcontroller unit. In the microcontroller unit C language coding is used for programming to control the robot. According to disease the injection mechanism work. Injection mechanism consists of three injections along with dc motor connection.

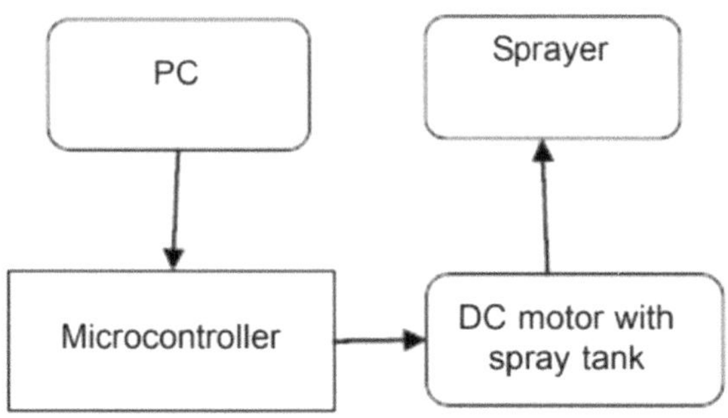

Fig.3. Spraying Techniques

These motor are used to take the corresponding pesticides in the injection when rotate in clockwise direction. When dc motor rotate in counter clockwise direction corresponding pesticides mix with water in tank.

Pesticides spraying are done through sprinkler motor. The control of spraying mechanism is done by the microcontroller unit. The DC motors are electronically controlled by microcontroller which receives input signals from RF module on the underside of the robot. By receiving the signal, DC motor is turned on and off to enable selective spraying of pesticides on plants. In this way DC motors help to spray the required pesticides on particular disease which is detected using image processing techniques.

MOBILE APP FOR PEST AND DISEASE MANAGEMENT OF CROPS

An app that allows farmers to identify pests and diseases using their mobile phones and provides remedial measures is the latest addition to using modern digital tools to benefit smallholder farmers.

Here are a few apps which you can download from the Play Store and App Store on your mobile phone to identify any plant diseases.

Agrio

Agrio is an AI-based plant disease identification app. The app uses images uploaded by users to match with data available in the app's database and then provides accurate details of the plant disease treatment.

Agrio has revolutionized crop monitoring technology by providing alters on users' phones using remote image sensing when any changes are seen in the crop field.

This app also helps farmers, and crop inspectors to record crop data on the go. Download for iPhone

CropsAI

An amazing app to monitor crop health. This app currently supports only a few crops and they are planning to include more crops in the future.

Using your mobile phone camera click pictures of the infested crop and upload them to CropsAI. With the help of AI, the app will give you the best advice to treat plant diseases.

Plants Disease Identification

Plants Disease Identification is easy to use. Just upload pictures of a leaf infested and the AI system of the app will detect the disease.

Currently supports a few crops, these are Apple, Blueberry, Cherry, Corn, Grape, Orange, Peach, Pepper, Potato, Raspberry, Soybean, Squash, Strawberry, and Tomato.

After the disease identification, you can contact your local agricultural expert for help. Download for iPhone

Plantix

This app is awesome. You just need to take a picture of the plant that is affected and upload it to the app. And Boom! The app will automatically identify the disease and give you all the details about the disease and its cure (biologically and chemically).

Plantix is now available in 18 regional languages of India. You can ask questions in the farmer community of the app in your own language and get answers quickly.

This app can guide you to plant a new crop through its pieces of information like when to plant, the best soil, when to water, etc. And also can help to calculate the amount of fertilizers to add to the soil. Overall a handy app for farmers.

Download for Android, not available for iPhone.

AUTOMATED DISEASE CLASSIFICATION IN AGRICULTURAL CROPS

Recently, the abundance of cost-effective smartphones among the farmers has created an opportunity for classifying diseases using the images of the infected foliage. The smartphone camera acts as a vision sensor which acquires color information in Red Green Blue (RGB) channels. This information along with the machine learning technologies learn to recognize the patterns of the disease based on the visible symptoms on the leaves. Traditionally features of the symptoms are represented using mathematical functions and evaluated in mapping for distinct patterns of different diseases. These handcrafted features are fed to the machine learning algorithms such as Neural Networks (NN), Support Vector Machine (SVM), etc., for classifying the different diseases and best features which result in better accuracy were utilized. Recently, due to the advancement in computing, a large number of artificial neurons are stacked in a specific architecture which forms deep neural networks and these are capable of learning the features automatically in contrary to the previous approach. These features are used for the image classification (in different domains) and this is popularly known as deep learning. One of the deep learning approaches namely Convolutional Neural Network (CNN) is widely used for image classification.

Different CNN based architectures namely AlexNet, VGG16, VGG19, GoogLeNet, ResNet, DenseNet, etc., have been developed and adopted for solving the problem of disease classification in various crops. Training from scratch approach requires a large dataset for training as it is a data-driven technique which poses a significant challenge to the researchers. Hence, an approach known as transfer learning is widely used where the architecture is pre-trained with the ImageNet dataset and each processing element has optimized weights as a result of training. These pre-trained networks have been retrained again with standard datasets. For example, PlantVillage includes images of common diseases from various crops using which many studies have been performed. The performance of these pre-trained networks is (generally) excellent and in some cases, it resulted in an accuracy greater than 99%. In few other studies, authors have used their own developed datasets for the classification of various crop diseases. Despite these efforts, the dataset for many crops is limited which is the critical input for training the deep learning-based algorithms.

This gap has motivated the authors to create a dataset for a few potential diseases in some vital crops (for which it is unavailable) based on the literature. Also, a rapid disease classification system with an easily available smartphone camera will be beneficial to the farmers. As an initial phase of this objective, a preliminary system consisting of a smartphone camera connected to a wireless Local Area Network (LAN) with a CPU will be used for classifying the disease with pre-trained deep learning architectures such as AlexNet, VGG16, VGG19, GoogLeNet, ResNet and DenseNet.

System overview

The system consists of a smartphone, wireless LAN and a Personal Computer (PC) (as shown in Fig 4). The smartphone used in the study was Lenovo A7700 equipped with 8 Megapixel camera which costs approximately U.S. $100. Wireless LAN is used for sending the acquired images from a smartphone to the PC.

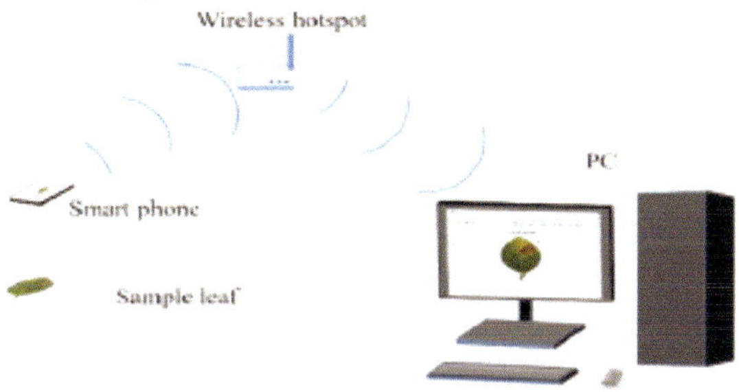

Fig. 4. Overview of the system.

A mobile application (IP webcam) has been used for acquiring and transmitting the images to the PC. The computer system used for the processing of images was ACER NITRO 5 SPIN laptop with 4 GB NVIDIA 1050GTX graphics card and 8 GB of Random Access Memory (RAM). The data communication and classification utilizing deep learning was performed using MATLAB 2018a.

Dataset and deep learning models

In this study, ten different diseases (as shown in Fig. 5) of four variety of crops namely eggplant (Solanum melongena), hyacinth beans (Dolichos lablab), ladies finger (Abelmoschus esculentus L) and lime (Citrus aurantifolia) have been considered for developing a disease classification.

Fig. 5. Disease selected for the study: (a) Tobacco Mosaic Virus TMV; (b) Little leaf disease; (c) Epilachna beetle; (d) Two-spotted spider mite; (e) Cercosopora leaf spot; (f) Brown spot; (g) Citrus hindu mite; (h) Citrus canker; (i) Yellow vein mosaic virus; (j) Leaf hopper.

The number of input images in the dataset has to be increased as deep learning models require a reasonably large dataset. Data augmentation is a standard procedure where the number of images was increased artificially by applying the image transformations such as rotation, translation and changes in intensity value which prevents a common deep learning problem of overfitting.

A random rotation angle within the limit of ± 30° was generated and applied to each image. In addition, a random number within the range of ± 30 pixels along the horizontal and vertical direction with a random intensity value in the range of ± 30 were used to transform each images.

These limits were selected based on numerous trials and the generated dataset was used for the training and validation process. Six pre-trained deep learning models namely AlexNet, VGG16, VGG19, GoogLeNet, ResNet101 and DenseNet201 have been used for training and validation of the created dataset. AlexNet, VGG16 and VGG19 is series CNN based network (as shown in Fig. 6) with repetitive convolution layer followed by nonlinear activation function namely Rectified Linear Unit (ReLU).

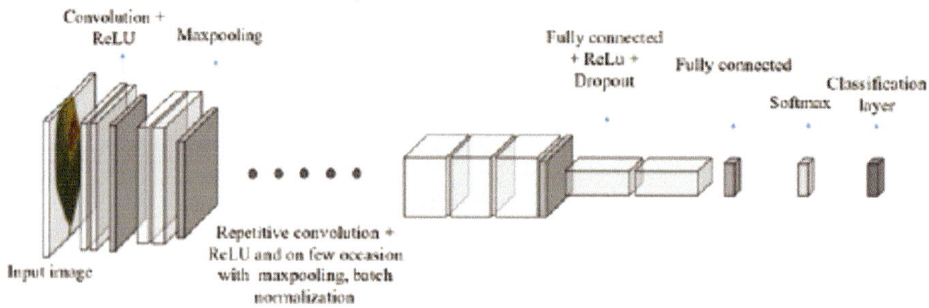

Fig. 6. A typical architecture of AlexNet, VGG16 and VGG19.

VISUALIZATION OF FEATURES

In a further sophisticated way, the learning of features by the trained model can be visualized and analysed qualitatively using deep dream image function (originally developed by Alexnder Mordvintsevl) available in the MATLAB software tool. Few studies have used visualization for understanding the learning ability of different architecture and convolution layers. Due to the limitation of the software tool used in this study, the function is applicable only for series networks (AlexNet, VGG16, VGG19) and it cannot be utilized for GoogLeNet as it is a Directed Acyclic Graph (DAG). The visualization gives an intuition on the learning ability of the network.

The function is applied to the last fully connected layer of AlexNet, VGG16, VGG19 as abstract features are combined by using the features from different layers. The above function generates an image which enhances the activation of the neurons in the network layers. The images were generated using the random image with pixels from a normal distribution. When the features were visualized with the AlexNet, it did not closely match the real symptoms. But, in the case of VGG16 and VGG19, features from symptoms of almost all diseases were able to be related to the visualization map. A sample visualization of Cercospora leaf spot is shown in Fig. 7, where random spots are present in the generated image.

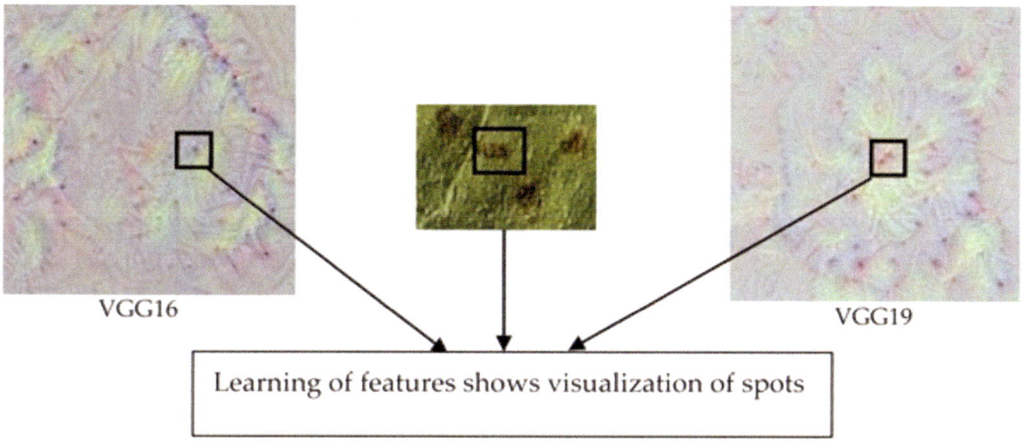

Fig. 7. Comparison of sample generated image with VGG16 and VGG19 using deep dream image of Cercospora leaf spot.

Testing

Finally, testing of the sample images obtained from the smartphone camera transmitted through the wireless LAN was performed. The images were read from the specific virtual COM port of the PC with the MATLAB tool. The testing was carried out in two ways, one with pre-processing and other without pre-processing. In pre-processing a binary mask was constructed using thresholding operation and an input test image was segmented with a background similar to the training and validation images. In the other case, the input image was directly classified with the trained model. The sample output obtained as a result of using pre-processed images for classification is shown in Fig. 8.

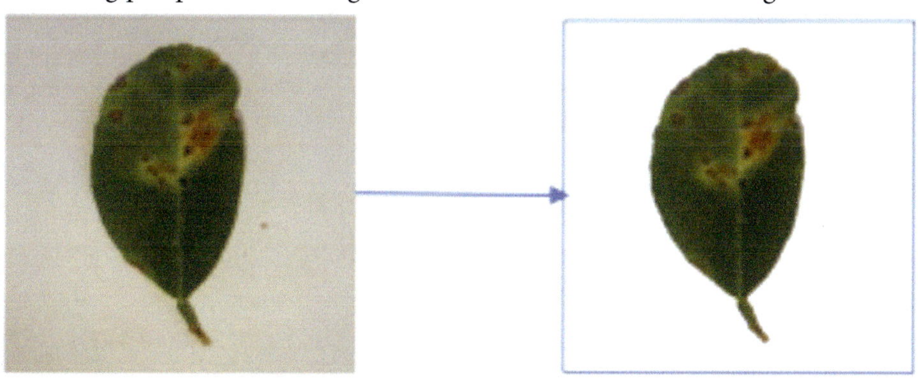

Fig. 8. Sample output result obtained using the pre-processed image of class citrus canker.

The prediction time of sample images for AlexNet, VGG16, VGG19, GoogLeNet, ResNet101 and DenseNet were 0.20, 0.39, 0.35, 0.27, 0.37 and 0.67 s respectively. In the test case, there was a significant difference in the classification accuracy with original and pre-processed images using different models. The applied segmentation procedure also resulted in the loss of few pixel information.

THE LEAF DISEASE DETECTION MODULE

The main objective of the leaf disease detection module is to identify the plant diseases using image processing principles with the help of CNN Algorithm. After identification of the disease, the module propounds the name of pesticide to be incorporated and the remedy to be followed so as to extricate the plant from further malady. Over the last few years, leaf disease detection has grown leaps and bounds when it comes to image disease detection. The deep learning algorithms have produced tremendous success in classification task. In this paper, the CNN Algorithms is trained to identify and classify the two most common endemics in plants: Bacterial Spot and Yellow Leaf Curl Virus.

The series of steps involves in disease detection module are as shown in the flowchart (Fig. 9)

AUTOMATION IN AGRICULTURE USING ARTIFICIAL INTELLIGENCE

With the advent of technology in this digital world, we humans have pushed our limit of the thinking process and are trying to coalesce normal brain with an artificial one. This continuing exploration gave birth to a whole new field Artificial intelligence. It is the process by which a human can make an intelligent machine. AI comes under the domain area of computer science which can be able to discern its milieu and should thrive to maximize the rate of success. AI should be able to do work based on past learning. Deep learning, CNN, ANN, Machine learning are certain domains which enhances the machine work and helps to develop a more advance technology.

Among all of these, the most widely used and constantly applied method for research purposes is ANN. Our human brain is the most complex part of the body. Based on the inter linked neural networks, electric signals traverses through the neurons with the help of axons. Synapses which are at the end of each node passes the signal ahead. ANN method was invented by keeping in mind the same concept of the working of the human brain. There are various algorithms of this approach such as for training this particular model algorithms like Silva and Almeida's algorithm, Delta-bar-delta, Rprop, The Dynamic Adaption algorithm, Quickprop are used based on its application. 9 neurons are used in the process. ANN is a task- based method which tells the system to operate based on some inbuilt task rather than a conventional computational programmed task. The architecture of ANN consists of three layers:

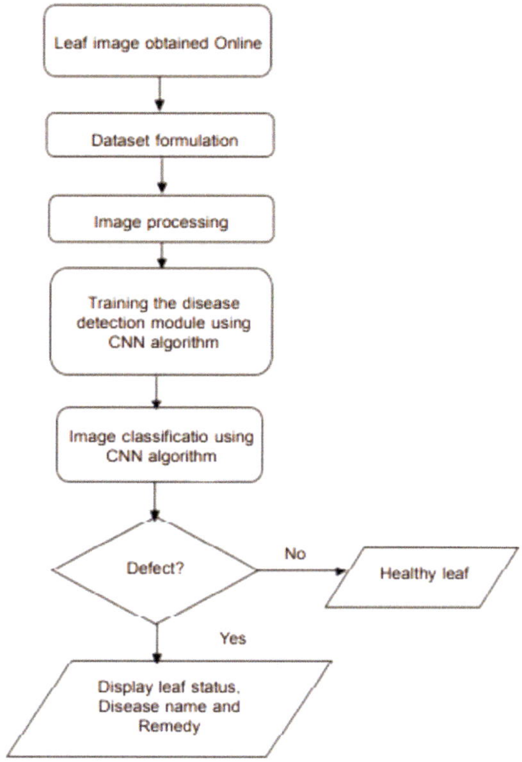

Fig. 9. Flowchart of Disease Detection Module

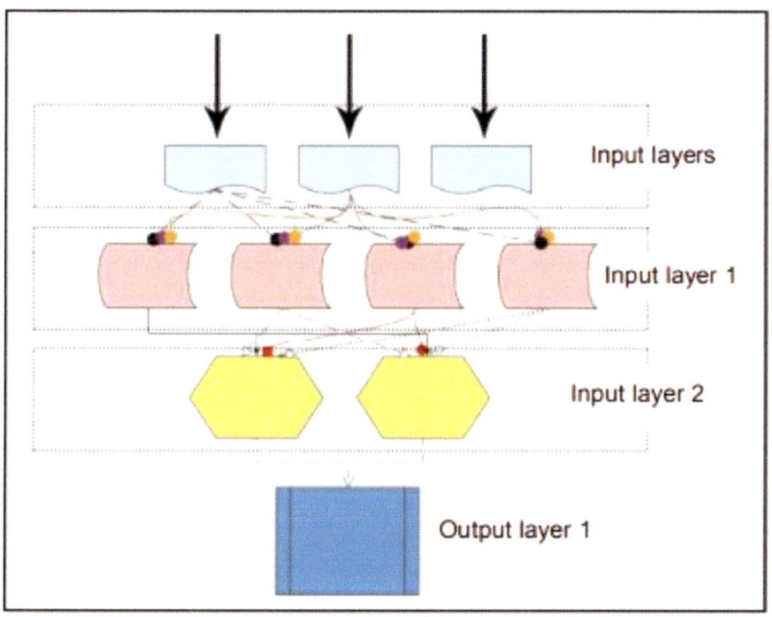

Fig. 10. Artificial neural network layers.

- Input layer
- Hidden (middle) layer
- Output layer (Fig. 10)

Feedforward back propagation mechanism and its parameters are shown above: Input Layer–7, output layer-1, hidden layer-50, number of iterations – 1200. Activation layer- Sigmoidal function in hidden and output layer, linear function in input layer.

Moreover, Artificial intelligence and machine learning are mostly hypothesis and theories. These are programming and algorithms. For the implementation of these algorithms and logic based concepts, there should be a hardware- software interface. The system through which this can be achieved is "Embedded systems". Embedded systems are hardware built systems consisting of memory chips with custom software programmed in it (Fig. 11).

ARTIFICIAL NEURAL NETWORKS IN AGRICULTURE

Artificial neural networks have been incorporated in the agriculture sector many times due to its advantages over traditional systems. The main benefit of neural networks is they can predict and forecast on the base of parallel reasoning. Instead of thoroughly programming, neural networks can be trained. Gliever and Slaughter (2001) used ANN to differentiate weeds from the crops. Maier and Dandy (2000) used neural networks for forecasting water resources variables.

Researchers brought together expert systems and Artificial neural networks in predicting nutrition level in the crop. Traditional ES (Expert systems) have considerable backdrops when it is being implemented. Use of ANN makes it up to all glitches of ES. The whole system is built on a single chip computer. Neural networks always prove to be the best when it comes to predicting methods. Neural networks can predict the complex mappings if a reliable set of variables are fed. To dodge the problems of frost formation in the fields, researchers developed a prediction model using neural networks. The model is first to feed with the raw data like humidity, temperature, precipitation, cloud cover, wind direction. Then, the data gathered got converted into binary data. These data, now are divided into two strings (input and output for the neural network model). The back-propagation network was used as a neural network predictor. A total of 10 trial sets were developed and trained by the model initially. The frost was predicted more efficiently when a range of values of parameters (mentioned above) was taken rather than single values.

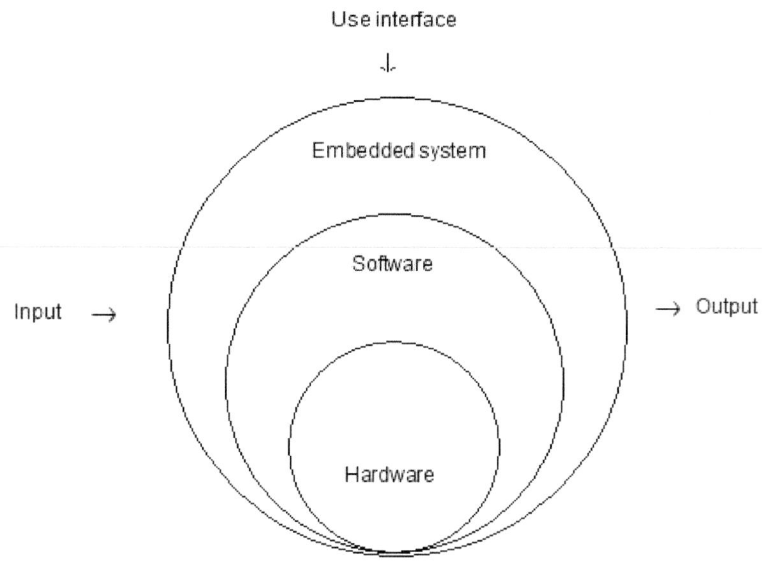

Fig. 11. Embedded systems.

Within the span of three years, two expert systems had been developed to increase the production of cotton crop. First, COMAX. In 1986, Lemmon made a successful attempt in developing an expert system called Comax (COtton Management eXpert). Lemmon, being the pioneer in AI in agriculture sector, develped a program called Gossym which is microcomputer friendly and bolster the use of Comax. For the first ever, the expert system (Comax) was successfully integrated with a computer model (Gossym) and simulated for the growth of cotton crops. This expert system was developed in order to operate continuosly throughout the year in cotton crop fields. Comax takes three parameters of the field into consideration; scheduling of irrigation, maintaining nitrogen content in the field, and growth in the cotton crop.

Second, COTFLEX. Another expert system for the cotton crop was developed. The system was named as COTFLEX. The system was made worked on Pyramid 90× computer which used UNIX as its operating system. The system incorporated the field and farm databases to provide important information regarding the cotton crop to the farmer so that it becomes easy for the grower to take critical and tactical decisions. The system developed in Texas, and it created simulation models and databases in the rule-based expert system to help Texan farmers take prudent economic and lucrative decisions. After successful testings, COTFLEX was imported to IBM microcomputer and was made open for the use.

Discusses about the soyabean crop growth model which is defined as SMARTSOY in the paper and the model is called SOYGRO. The model is explained by knowledge based approach where it is divided into two approaches first being the positivistic approach stating the attempts to duplicate the processes of domain experts in order to come to a conclusion while the second approach being the normative approach which attempts to duplicate the conclusions excluding the processes of the domain experts. The damages caused by insects are determined by systematic method for determining the damage rates and the cost control. Here, the positivistic and normative approach collapses because the later method helps it the selection of insecticide and application rate. However, the systematic method does not help to find the insect damage rate on yield. This is a major drawback for generating recommendation because we need to mix both the approach as the yield reduction is calculated by previous experiences of the similar insect attack, pesticides used and the outcomes of the crops in the end. The goal to generate recommendation specifically for soyabean crop is based on calculation of the damage rate and the cost to treat the plant and gain the yield. This calculation is derived by both the approaches.

An expert system PRITHVI based on fuzzy logic was developed in Rajasthan, India. The system was designed explicitly for Soybeans crop. This system gathered its knowledge base from agricultural officers, published literature, and experts of soybean crops. Fuzzy logic was considered in studying the whole system and advising the farmer as an expert. PRITHVI was divided into five modules. The main aim of developing this expert system was to help the farmers in the region increase their soybean production. The system used MATLAB as a user interface modul

Researchers developed an expert system which helped the farmers with when to spray insectisides on the apple fruit to avoid the damage due to insects and atmosphere. The system was named as POMME. Along with the time, it also advised the farmers what to spray. Instead of theoretical values from the infection table, here apple scab disease cycle model was used in POMME. The results of the system were satisfactory and system was approved by the experts who had used it on trial basis.

A prediction model was developed. As mentioned above, the prediction model of this system had three layers. The efficiency of the model was dependent on the number of the hidden layers. First and foremost, the ANN model was built and trained using various algorithms such as Silva and Almeida's algorithms, Delta-bar-delta, Rprop, and various other to find the most favourable configuration. Trial and error method was implemented to choose the number of hidden layers. There should be a précised way to scrutinize the selection of some hidden layers because the prediction system's accuracy is dependent on the number of hidden layers. It was observed in the research that more the number of hidden layers in the ANN model; the more accurate were the prediction.

Since the purpose of the system was to make it handy for the farmers, it is developed on APK platform. The source code was written in Eclipse with Java codes in the backhand, and the algorithm was developed using Matlab and ANN toolbox. The whole file was then extracted on the Android platform so that it can be utilized by smartphones. Besides suggesting the crop to the farmer, the system also has the additional advantage of advising the farmer for the fertilizer to be used if the farmer wishes to use the crop of his choice.

Evapotranspiration process is imperative for maintaining the stability in the hydrologic cycle, sustainable irrigation method, and water management. Parameters- Elevation, Mean daily Temperature, Max. daily temperature, Min. daily temperature, Wind Speed, Relatuve Humidity, Sunshine Hours, Daylight hours, Latitude, Condition coefficient. There are more than 20 established method to determine ET which is dependent on several parameters.

A profound study was carried out in the valley of Dehradun; India was assessing the importance of the addition of ANN in several techniques of estimation of ET. Researchers gathered monthly climate data from the Forest research institute (FRI) Dehradun for ET estimation. The methods on which the algorithms were applied were: 1. Penman-Monteith method 2. Levenberg-Marquardt back propagation. It was observed that increasing the number of hidden layers in the system resulted in instability in the ET estimation. So, training function with optimum trial and error method is to be chosen for the overall optimized estimation of ET. It was observed that out of six training algorithms of ANN model, function training with 75% data feed in it was precise and had the best number of neurons. Furthermore, there was a gauging between PM method and ANN model with the single layer feed forward back propagation algorithm. ANN model was designed and developed using Matlab. Six algorithms were conjured and assessed. As evapotranspiration is of vital importance in irrigation and water management, this research manifested the predictive prowess of ANN structure if implemented correctly.

Furthermore, also a method was created to discriminate weed from crops with the help of image analysis and neural networks. It had the accuracy of more than 75% without any beforehand plant information fed into the system. Researchers developed expert systems based smart agriculture system. The concept of IoT in this system was to send the data to the server so that actuators of the field should be able to take appropriate decisions. For that, the server should be intelligent enough to take decisions independently. This system consists of temperature, humidity, leaf wetness, and soil sensors. It only gives the information about the field and doesn't act on the irrigation process.

Researchers developed two ANN models to estimate soil moisture in Paddy fields using decidedly less meteorological data. Both these models were then corroborated and validated by studying observed and estimated soil moisture values. The first ANN model was generated to get the estimate ET. The help of minimum, average, and the maximum air temperature was taken. To develop the second model, solar radiation, precipitation, and air temperature data was gathered. Both these models resulted in the accurate and reliable estimation of soil moisture in the paddy fields by using the least meteorological data, less labor and time consumption.

Researchers discuss the neuro drip irrigation systems where ANNs were developed to predict the spatial water distribution in the subsurface. For drip irrigation method to properly function, water distribution in the lower level of the soil is of the grave importance. Here, ANNs makes the prediction which comes handy for the user which in turn results in the fast decision-making process. ANN models give the result of wetting patterns (first and second) after the soil is infiltrated with the water from the emitter which is on the surface of the land. Thus, the ANN model provides continuous patterns to the user. Also, researchers developed a model to study the yield of the maize crop. A multi-layered feedforward ANN (MLFANN) is used. To fuel such network, learning algorithms like GDA (gradient descent algorithms) and CGDA (Conjugate gradient descent algorithm) are employed. Both the algorithms have been written and simulated in the MATLAB using nueral network toolbox.

Precision agriculture and WSN applications combine an exciting new area of research that will greatly improve quality in agricultural production, precision irrigation and will have dramatic reduction in cost needed. Furthermore, the ease of deployment and system maintenance, monitoring opens the way for the acceptation of WSN systems in precision agriculture. Using the proposed methodology, in finding the optimal sensor topology, we contrive to lower implementation cost and thus make WSN a more appealing solution for all kinds of fields and cultivations.

AUTOMATION AND WIRELESS SYSTEM NETWORKS IN AGRICULTURE

It is imperative for any sector to evolve with time. The agriculture sector had to adapt the breakthroughs and inventions which came along in automation field. Researchers came forward with emerging research area of embedded intelligence (EI). Embedded intelligence in agriculture sector includes smart farming, smart crop management, smart irrigation and smart greenhouses. It is necessary for a nation to include these growing technologies in agriculture sector for growth of a nation as many sectors are inter-dependent on agriculture. Also, researchers of this paper demonstrated Technology roadmap (TRM) which in turn clarifies the qualms regarding the areas of agriculture mentioned above (smart farming, smart irrigation etc).

Taking into consideration the socio and economic vitality of agriculture in India, researchers Patil and Thorat (2016) developed a system which predicted grape disease beforehand. Any anomaly in the grape plant was noticed only after it was infected and this had a considerable deteriorating effect on the whole vineyard. The system employed various sensors such as temperature sensor, leaf wetness sensors, and humidity sensors in the vineyard. These sensors send the data sensed to the database in the ZigBee server which is linked to the sensors. Deployment of Wireless System Network (WSN) in any field needs to satisfy certain criteria and Zigbee alliance has developed open global standards called ZIGBEE. Zigbee compliances of four layers namely physical layer, medium access control layer, network layer, application layer as stated.

The three devices: Zigbee Co-coordinator (ZC), Zigbee Router(ZR), Zigbee End Device(ZED) has different function in the WSN. Researchers discusses end to end approach of Zigbee in agriculture. The server will store the data. The server is commissioned with a hidden Markov model algorithm in it. This algorithm is present to train the normal data sensed by the sensors and report any aberration in temperature, humidity or leaf wetness which can result in grape disease to farmer via SMS.

Machine learning is blended in the system beforehand for astute deduction of disease in grapes. The additional advantage of this system is it also suggests the farmer pesticides and pacifies manual effort in the detection of disease. While a similar method of machine learning was employed in monitoring the growth of Paddy crops. This system was developed for increasing the yield and productivity of paddy crops. It also proved to be cost effective and durable.

The sensors used in Fig. 11 are for monitoring agricultural field are shown in Fig. 12. Sensors such as MQ4 and MQ7 are used for Natural Gas sensing and Carbon Monoxide sensing respectively. DHT11 is used for Temperature and Humidity monitoring of the environment, soil moisture sensor is used for measuring soil moisture level and have continuous monitoring. Esp8266 is a wifi module which helps in communication between the hardware system and the device which users use.

Use of wireless communication has changed the standards of communication in today's world and this can also raise the standards of agriculture automation. Researchers has bifurcated the IOT gateway into different nodes such as actuator, sensor, interface and wireless link which give assistance to communication between them. Frequency estimation and the bandwidth requirement for this communication has also been elaborated which can be very beneficial for automation.

Many different IEEE standards describe sensor networks, standards like IEEE 802.15.1 PAN/Bluetooth, IEEE 802.15.4 ZigBee and many more are necessary to know while planning its application. Researchers also discussed about IPV6 the Internet Protocol for wireless communication and also many hardware system for establishing a WSN. By using WSN, Precision farming is possible and the strategy is used for crop management. Different data is recorded by the sensors and stored in the system. The system is made to learn by the previous data from the sensors and future actions are taken accordingly.

Researchers studied three soil moisture sensors in Peacan crop field and came to a conclusion that the sensors used (ECH_2O_{-5}TE, Watermark 200SS and Tensiometer model R) need site specific calibration in order to bring accurate results.

The highlighting features of the paper presented by researchers includes smart GPS based remote controlled robot to perform tasks like; weeding, spraying, moisture sensing, bird and animal scaring, keeping vigilance, etc. Secondly, it includes smart irrigation with smart control based on real time field data. Thirdly, smart warehouse management which includes; temperature maintenance, humidity maintenance and theft detection in the warehouse. Controlling of all these operations will be through any remote smart device or computer connected to Internet and the operations will be performed by interfacing sensors, Wi-Fi or ZigBee modules, camera and actuators with micro-controller and raspberry pi.

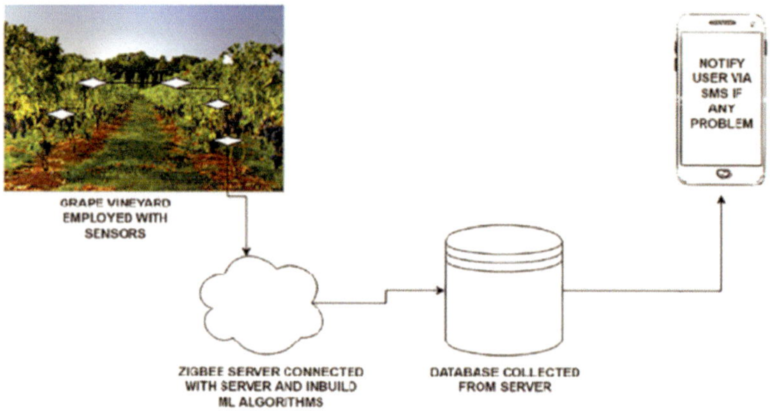

Fig. 11. Grape disease detection system using ML algorithms.

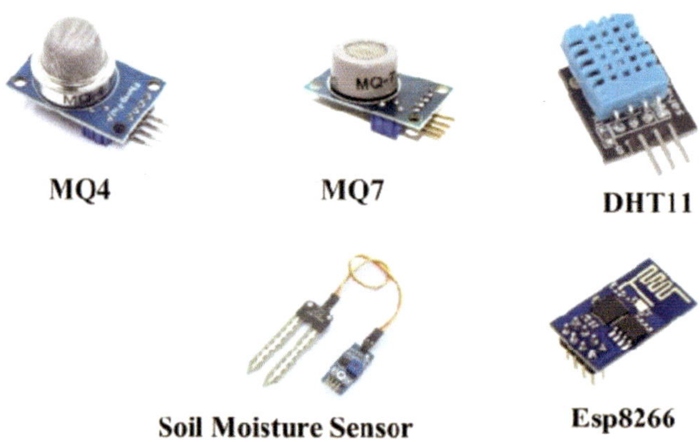

Fig. 12. Different type of sensors used for detection.

Thermal Imaging is a noncontact and nonintrusive technique which analysis the surface temperature of the agricultural field and provides valuable feedback to the farmer. Researchers discussed the use of cloud based thermal imaging system which helps the irrigation by incorporating the performance of the equipment's and determine the area of field which requires the water most. The absence of uniformity will hamper the crop growth and thermal imaging can help to consolidate this loss. Also, Thermal Imaging is put to use in agriculture sector truly because of its wide application.

Researchers discussed the use of robot in the agriculture field. The robot is designed to follow the track of white line where actually there is a need to work and other surface is considered as black or brown. Working of robot is for spraying of pesticide, dropping of seed's, water supply and ploughing. In 2016 a group of researchers came up with e-Agriculture Application based on the framework consisting of KM-Knowledge base and Monitoring modules.The systems developed in IOT and Cloud Computing emphasizes on reliable architectures to provide timely information from the field over 3G or Wi-Fi.

TI CC 3200 (RFID) launchpad was used to build the prototype with other necessary devices. Knowledge base has advantage over conventional IOT based systems; Knowledge Base is constructed to store vast complex structured and unstructured information to assist farmers or even an individual with no prior knowledge of farming. But finding right information in an appropriate manner is difficult where providing relevant knowledge should be distributed not only in an organised and complete manner, but also in absolute way. The knowledge based infrastructure allows adapting the changes in agriculture for a better extension and adding advisory services.

As degree of automation is required in each and every field so the human intervention becomes less and it is very important to design a layout in the early stages of the mechanics and electronics. Weed management is the issue which farmers face a lot and computer vision can help to solve the issue. There is particular difference between a weed and the desired crop. CNN can help to distinguish among them and notify us to cut only the unnecessary plant. CNN has many algorithms which can even be used to identify plants and get the data accordingly for plantation.

R-CNN extensively used in object detection and in automation it is used for fruit detection and counting of fruits. Researchers discusses the use of R-CNN in fruit detection of orchards, while training the input to the network is 3 channel colour image (BGR) of arbitrary size. They have used VGG16 NET with 13 convolutional network and also ZF network which has 5 convolutional layers. Data augmentation is used because it helps in artificially enlarging the dataset and changing the variability of the training data. The results discussed by them are promising as for both mangoes and apples Faster R-CNN outperformed the ZF network approach. (R-CNN stands for regional convolutional neural network).

Cloud based decisions and support in the agriculture is booming now a days. The Decision support and Automation system (DSAS) helps the farmers of the growers to control all the applications through its web portal. DSAS as different stages where the it can interconnect many devices on the single time and give the real time data to the farmer. The farmer plays the vital role as he can monitor the real time data and also control all the machine through software's. Systems like spray controller will spray the pesticide on the field in a defined amount. Similarly, irrigation controller helps to manage irrigation and fertilizer controller manages fertilizer. DSAS works through the data given by different sensors like soil moisture sensor, nitrogen sensor, etc.

IMPLEMENTATION OF FUZZY LOGIC SYSTEMS IN AGRICULTURE

Researchers developed an FK-based fuzzy model to decide the land suitability. Various fuzzy sets were generated using farmer's as well as scientific knowledge congruently. The sets used S-membership functions and were used to determine soil texture, slope, and colour. The research work was done in several villages of Nizamabad district of Andhra Pradesh state of India. In FK-based fuzzy factor maps, it is not necessary to take the lower and upper limits 0 and 1 respectively. This is because in knowledge-driven fuzzy modeling there is no constraint on choosing the membership functions as long as the functions are in context of the factor which has to be modeled.

Another implementation of fuzzy modeling was done for land levelling by researchers. They employed fuzzy control theory in the controller of the system. By implementing fuzzy control theory, a precision based result was obtained. High accuracy fuzzy control theory translates the variables (the deviation in the height of the field and expected the height of the field) into the defined variables sets (E and EC) which contains fuzzy terminologies such as 'High', 'Very high' and so on. There are nine sets defined for variable set E and two for variable set EC.

This theory helps the controller deduce the position of the bucket which is inturned will be the height of the field. The bucket receives the signal from the receiver. Researchers developed an innovative system for grading the leaf diseases. The system was segregated into five parts namely Image acquisition where the researchers have captured images of Pomegranate leaves, image pre-processing where the captured image is then resized, filtered, and processed according to the required parameter. Then comes colour image segmentation where k-means clustering is used to isolate the healthy part of leave with the disease infected part. Afterward from the resized image, total leaf area is calculated, and with the help of the third part, infected disease area of the leaf is calculated. Finally, in the last part, with the help of a fuzzy inference system, accurate grading of the disease can be taken out.

FIS (Fuzzy inference system) was developed by researchers to determine optimum rates of N fertilizer on the basis of field and crop features. Also, researchers implemented FIS to estimate stem water potential. Kavdir and Guyer (2003) applied FL model in determining the quality of Apple fruit. Gottschalk et al. (2003) developed fuzzy logic based air controllers to maintain the temperature of storage facilities for Potato. Researchers came up with a simulation software (SCD) which came in handy for many fuzzy based controllers. The software used rule-based knowledge base with IF. THEN condition type. Its graphical characteristics make the software adaptable to any fuzzy algorithm simulations. Another Fuzzy inference system using IF...THEN condition type was developed. The model forecasted plant disease on the base of weather data. The system was developed to avoid diseases in plant beforehand as disease occurs in specific range of temperature and humidity in the weather.

India and China alone constitute 2.7 billion people living under the stress of water shortage. Out of overall water consumption, 70% is consumed in the agricultural process. Remaining is used in infrastructural pipelines and other miscellaneous works. Water leakage is inevitable and uncontrollable in cities. Water demand will shoot up by 50% shortly and this fact cannot be vetoed away. The farming fraternity is the only option in which water usage can be optimized by employing smart irrigation systems. By inculcating smart irrigation system, wastage of water can culminate to a great extent can abridge water consumption by 20%.

There is a major problem of water wastage and a dearth of water in conventional irrigation methods employed. To give an example, Egypt faces a problem of water distribution from the Nile river with neighboring countries. Ample research had been carried out to solve the problems faced in the irrigation, process. Many companies have developed a sensor-based smart irrigation system. These systems have been developed for optimal water usage, monitoring of water pollution, and to take care of some other grave problems. Soil moisture and temperature sensors interact directly with embedded components in the field and take care of required water distribution among crops without farmer's interaction.

Water which is to be fed to the farms, either by the means of smart irrigation or any other conventional method, should be of a good quality. Researchers have started implementing IOT systems and Artificial intelligence techniques in aquaculture sector along with agriculture.

Researchers came up with a smart system which controlled valves of sprinklers with the help of temperature and moisture sensors deployed in the field. However, this system did not consider the water pollution problem. Researchers came up with a distributed irrigation system which works on soil water measurement. M2M (machine-to-machine) technology which allows machines to interact with each other autonomously and store the data directly in a cloud-based server online. This M2M technology is in an incipient stage and is developing steadfastly. Researchers developed a technology which allows machines to communicate themselves. Researchers also developed a complete sensor-based intensive irrigation method which is self-organizing. This system constructed a bottom and upper layer. Researchers tried to demonstrate a prototype of the small-scale smart irrigation system.

So far, there hasn't been such advent which allows complete freedom of human intervention. This paper tries to bring forward a method through which with the help of AI and embedded technology which eliminates the glitches emphasized in the past.

FUTURE SCOPE

The farmers who are young will make more investments in automation with much interest than the elder farmers. The technology which is new has to be introduced slowly with time. Slowly the agriculture sector is moving towards precision farming in which management will we done on the basis of individual plant. Deep learning and other extend methods are used to detect the plant or flower type, this will help farmers to provide favourable environment to the plant for sustainable growth.

Eventually the production of more customised fruits and plants will grow, which leads to an increase in the diversity of products and production method. Artificial intelligence techniques are growing at a rapid scale and it can be used to detect disease of plants or any unwanted weed in the farm by using CNN, RNN or any other computational network.

Greenhouse farming can provide a particular environment to the plants but it is not possible without human intervention. Here, wireless technology and IOT comes in the run and using the latest communication protocols and sensors we can implement weather monitoring and control without human presence in the farm. Harvesting of fruits and crops can also be incorporated by robots which are specialized in working round the clock for quick harvesting.

Application of robotics are vast in farming such as the robots can be used in seeding and planting, fertilizing and irrigation, crop weeding and spraying, harvesting and shepherding. To complete the same work in many cases, it would take approximately 25 to 30 workers.

Thermal Imaging can also be implemented by using drones and thermal camera in it. The drones monitor the farm and gives continuous real time data of the field so that the farmers could know in which area of the field the water quantity is less and can only start irrigation in that particular area. This will prevent water flooding or scarcity of water in the field and the crops get advent amount of water all the time. Many different integrated approaches can be used to provide a viable environment and increased growth.

PLANT-DISEASE-DETECTING ROBOT

A smart robot with disease detection, plant monitoring and labour-saving capabilities is in the works for Ontario's green-house vegetable sector. The Guelph Intelligent Green-house Automation System (GIGAS) is an integrated robotics system under development by the University of Guelph to automate some of the most labour intensive greenhouse tasks.

The system had its start as a robot to harvest produce and remove leaves from plants. An early prototype was armed with visioning technology that allowed it to identify produce ripe for harvest and a special gripper that let it pluck fruit from the vines without damaging it.

His current prototype is a greenhouse scouting robot that searches for early signs of plant disease. Not only can the robot detect symptoms not yet visible to humans, but it can be scouting constantly whereas a grower might only get to every spot in a greenhouse once every week or two.

GREENHOUSE MONITORING ROBOTIC SYSTEM

Description of working environment

Fig. 13:The disease-sensing robot that will scout greenhouses

Commercial vegetable greenhouses in northern climates are designed to create suitable conditions for growing and protecting plants in cold weather. They are built with a steel framework and covered in clear polyethylene film, polycarbonate, or glass. Although greenhouses control their climate, outdoor conditions still have a considerable impact on the internal environment. Most notably, indoor illumination of plants varies significantly from day-to-day and plant-to-plant as solar angles and cloud cover change. In order to maximize yield, plant density is kept very high. Each hectare of a tomato greenhouse typically contains 24 thousand plants, and the size of the average greenhouses in Ontario is three hectares. Plants are grown every half meter along double rows, each row about a meter wide and 60 100 m long. Over the course of the growing season, plants reach heights of up to3.5 m, and as they grow taller they also grow closer together, coming into contact with plants from adjacent rows. Greenhouses use land much more efficiently than open fields, however, they still require a considerable amount of real estate which forces them away from larger cities. The greenhouses used for testing in this study were approximately 300 km away from the research institute and two hectares (five acres) in size.

As tomato plants mature, their lower sections are de-leafed to promote fruit development and to prevent disease. Figure 14 shows the view of young and mature plants from within a row. In order to maintain adequate separation between plants and to promote vertical growth, plants are bound by string and attached to the overhead metal framework. As their fruit matures, each plant is manually lowered allowing for a new set of flowers to develop at the top of the canopy and for a ripe fruit to be harvested near the bottom.

Water filled pipe-rails are placed between rows to help control greenhouse temperature and for use as a track for rail transport of service vehicles. In a two-hectare greenhouse, there are between 120-200 rows with a combined length of 12 km. Greenhouses are typically split in half by a wide concrete walkway for accessing rows. Figure 15 shows the inside of a greenhouse from the perspective of looking down the center walkway and down a single row.

Because rows are dead-ended the effective walk-through distance of a two-hectare greenhouse doubles to 24 km.

Figures 16 and 17 illustrate a top-down view of the greenhouse pipe-rail arrangement.

Tomato greenhouse temperatures are targeted to be between 18 - 28 °C with a humidity between 60 - 70% RH, although outdoor weather conditions often interfere with the climate control system and cause the values to fall outside of their optimal ranges. During periods of heavy rain, leaks can develop in sections of the roof. Shadows produced by clouds, infrastructure, or other plants can vary illuminance from 110 000 lux. Ventilation issues caused by faulty fans and ceiling vents can cause significant differences in temperatures and humidity, and roof structure causes shadowing to change throughout the day. Constant deviations from optimal conditions in localized areas can impact the uniform development of fruit, and produce environments more favorable for disease development.

Fig 14: View of tomato plants from within the row taken perpendicular to row length. The appearances of tomato plants vary significantly between the beginning of the growing season (left), and their fruit production phase (right).

Labor operations take place during daylight hours with 14 to 40 people working in a two-hectare greenhouse. The number of workers changes depending on the stage of crop development, harvesting being the most labor-intensive period. Workers spend the majority of their time within rows, often on manual or powered rail vehicles. Some greenhouses have labor monitoring stations placed at the heads of each row for workers to record their current location and activity.

Fig. 15: Typical commercial tomato greenhouse environment. The left image shows a perspective looking down a 4.4m wide center walkway. The right image is a view down a row, with pipe-rails visible near the ground.

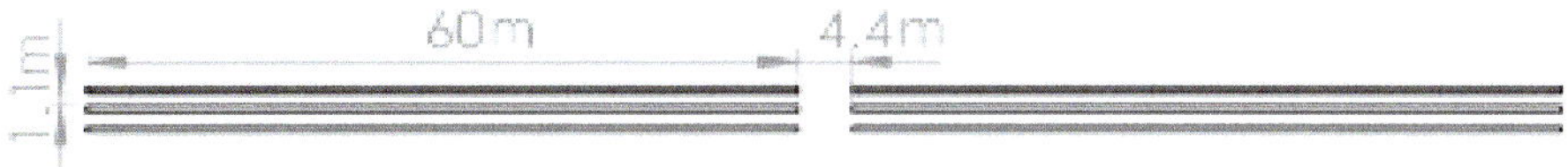

Fig. 16: Top-down view of a typical tomato greenhouse pipe heating and rail system. The walkway between rows is 4.4m wide, and the distance between rows where plants are grown is 1.1m.

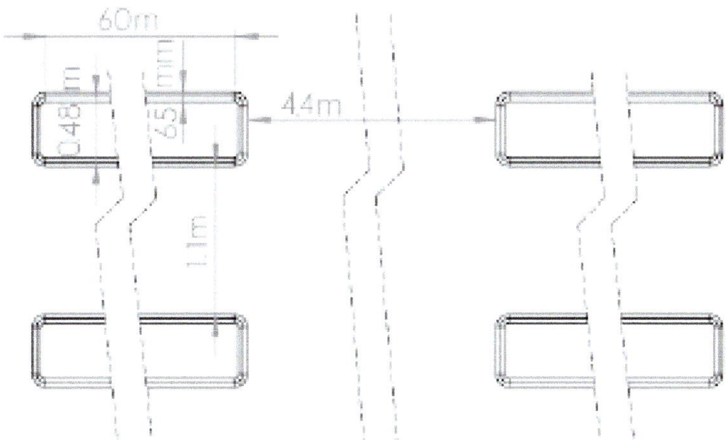

Fig. 17: Detailed top-down configuration view of a typical tomato greenhouse pipe heating and rail system. Pipe diameters and distances between rails can vary considerably between greenhouses.

DESIGN CHALLENGES

Greenhouses have high capital costs and as such, it is not feasible to request major changes to existing infrastructure to accommodate a new robot design.

- Integration with existing greenhouse structural, environmental, and labor conditions is essential for practical implementation
- Extensive clutter, occlusion, and uncertain nature of plant growth
- Safe operation for human laborers and plants

Robot architecture

The primary requirement of the plant monitoring robot is to autonomously collect daily imaging for each plant. Ultimately, the collected data will be used to develop machine learning models, generate recommendations for growing strategies, and present images augmented with additional insights to growers for further analysis. Here we identify the following functional and non-functional requirements.

Functional requirements

- Collect and store image data of every plant once per day
- Transmit collected data to remote system
- Autonomously recharge batteries
- Provide manual control for ad hoc repositioning by personnel
- Provide user control and feedback for scouting operations
- Capture objects of interest in images at sizes suitable for recognition
- Autonomously navigate within and between rows, transferring between concrete and pipe-rail
- Carry a payload of at least 100 kg
- Run continuously at 3.6 km/h for a minimum of eight hours before recharging
- Provide 250 W of continuous and 1000 W of peak power
- Protect sensitive components from liquids, condensing humidity, and temperatures over 50 degrees Celsius

Non-functional requirements

- Be safe to operate and avoid damage to personnel, crops, equipment, and infrastructure

Hardware design

Mechanical platforms

The prevailing mechanical design of crewed ground vehicles, including passenger cars, uses a single set of wheels in an Ackermann steering geometry, Figure 18. This configuration is suitable for long-distance travel along a relativity uniform surface, including paved roadways, mild off-road terrain, and factory floors. Un-crewed ground vehicles generally share the single wheel-type design of crewed counterparts but more often use differential steering for simplified control and greater manoeuvrability. This alternative steering geometry comes at the cost of reduced power efficiency and increased tire wear.

Railway vehicles generally have no active steering mechanisms and instead use steel, conical wheels with a rigid axil to passively follow track layout. Rail travel is more energy-efficient and is simpler to navigate than road travel, but has reduced flexibility. In order to benefit from the flexibility of roads and the efficiency of rails, road-rail vehicles have been developed, but have not been adopted into general wide use. However, for specialized applications including railway maintenance and greenhouse harvesting, road-rail service vehicles are commonly used to accommodate the requirements of traveling on both paved road and rails by using a dual wheel-type configuration.

Mk I mechanical platform

The first automated plant monitoring prototype used a base that was designed and built by South Essex Fabricating (SEF) Inc. in Leamington, Ontario. The vehicle is used for moving through rows on rails in the commercial greenhouses visited during experiments. It is typically equipped with a pneumatically powered elevating work platform used by operators to access the full height of the canopy, up to 4 m. The vehicle was modified to not include the elevating system when used as the base for the Mk I, reducing its weight by more than 100 kg and decreasing the chance of toppling. The maximum speed of the SEF rail transport vehicle is 3.6 km/h. Figure 19 shows a modified version of the vehicle without an elevating work platform.

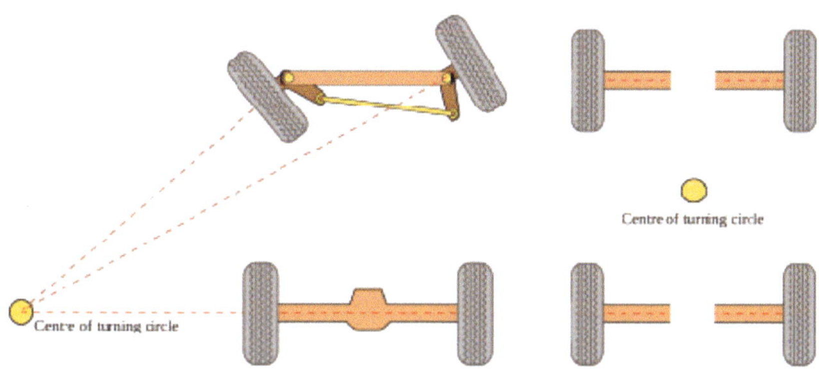

Fig. 18: Ackermann (left) and differential (right) steering geometries with a single set of wheels. In the differential arrangement, the speed and direction of each wheel is powered and independent.

The vehicle was not designed for motorized travel on paved roadway, and as a result, its primary wheels are not well aligned and have no steering capabilities. Instead, the vehicle includes a pair of unpowered pivot wheels to balance the vehicle in the center.

The pivot wheels are lowered with pneumatic actuators to form a seesaw, raising either the front or back wheels, and allowing the operator to turn the vehicle about the pivot point. This design provides the same maneuverability as a differential steering system. Due to poor wheel alignment, the pivot wheels must be used to move the vehicle when on concrete, even in straight lines.

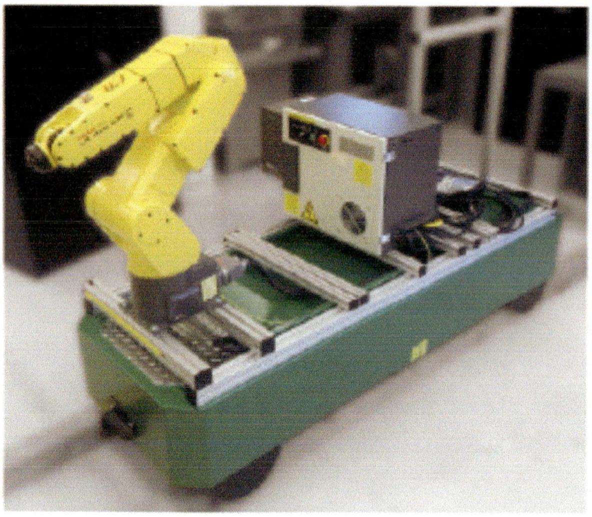

Fig. 19: South Essex Fabricating (SEF) Inc. service vehicle, used as the base for the Mk I plant monitoring prototype. Shown here with scissor lift replaced with custom mounting rails and a robotic arm and controller.

A panel is normally mounted on the platform to provide control for direction, throttle, and position of the pneumatic pivot wheels. There are two safety mechanisms built into the vehicle.

A foot pedal in the middle of the platform must be depressed to activate the throttle, and a rail sensor prevents lowering of the pivot wheels during on-pipe travel, as this would result in derailment.

Safety logic is built using electrical relays. Power to the motor and actuators was provided by two 12 V rechargeable leadacid gel batteries connected in series for a 24V power rail with 110 Ah. Batteries were charged using a built-in 110 VAC adapter with a standard North American mains power connector.

While this platform was capable of traveling autonomously in each row, the nature of the steering mechanisms required significant physical force to maneuver the vehicle off-rail and prevented the automation of greenhouse row transfers.

Mk II mechanical platform

The second automated plant monitor prototype used a platform based on the Dutch built Metazet Formflex M-Service vehicle, designed for personnel transport in commercial greenhouses for use on both pipe rail and concrete passageways. This vehicle has a flatbed standing area without an elevating platform. Large rubber-filled wheels are used when traveling on the concrete walkway, with smaller steel rollers attached to a shared axle for travel along pipe rails. The steel rail rollers can be adjusted to fit a range of pipe rail spacing.

12

AUTOMATION IN AGRICULTURAL EQUIPMENT AND TECHNOLOGY

The agriculture industry in the modern era is developing by leaps and bounds. The rise in the global population has caused an increase in food demands which in turn incurs farming produce price increase, reconsidering trade policies, shifting supply chains, and reshaping traditional food delivery routes. Despite these challenges, the amount of agricultural workers is steadily declining. Therefore the farmers are facing a critical juncture between wanting to produce more, higher-quality crops and finding the workers to plant, maintain, and harvest those crops. The necessity arose to focus on using technology to produce more food in order to meet the demand and the concept "Smart Farming" came into place which makes farm more efficient and automates the crop production cycle.

In the modern world there are numerous technologies that are commonly being utilized in agriculture. The advent of Agricultural robots simply called as agribots was a turning point in the field of agriculture that has resulted in producing more and higher quality foods with less manpoawer.

You can easily find a modern farmer spending more time looking up into the sky rather than gazing at the crops and soil below. No, they are not marveling at the blue scene with puffy clouds, full of promise for an ample harvest. They are watching the drones flying over their land and making sure their crops are safe and sound from tilling to harvesting. The ubiquitous implementation of custom software-powered agricultural drones is a small part of the farm automation practices that have been set in motion by the challenges today's farmers face. Modern problems require modern solutions, right? The shift to automated farming is inevitable, and believe it or not, it's not about cutting jobs. It's about achieving better results as well as safer and more comfortable working conditions.

Monitoring of the agricultural environment has become a major issue in recent years due to a range of factors including population growth, need for increased food production and the apparent onset of global warming. Population growth has led to environmental problems arising primarily from the pressure to produce more food from an essentially static or even diminishing food production land area. Modern agriculture depends heavily on engineering, technology, biological and physical sciences. Mechanization of agriculture activities have relieved much of the manual work and increased efficiency and productivity of farms.

AUTOMATION IN FARMING

Two primary technologies stand behind farming automation: machine vision and IoT. Backed by GPS and GIS, which guide their movement in the fields, robotic farm automation solutions can create a virtual 3D model with the help of high-resolution cameras and sensors, which then help them navigate freely and perform tasks seamlessly.

Now, not all agricultural robots are fully automated. Remember: they are robotic solutions and not robots that have come here to make Megathron prevail over the Earthlings. Agricultural robots are programmed to perform various levels of automation, which makes crop management software play a crucial role in the configuration of automated farming robotics.

Humans will have to learn how to manipulate these robotic tractors, which on the one hand is quite a tall order. On the other hand, it renders agricultural employees' lives easier, safer, and more comfortable. Navigating a tractor while sitting in a soft chair with a tablet in front of you is way more pleasant than going out into scorching fields in skin-drenching winds.

AGRICULTURAL MECHANIZATION STRATEGY

Definitions

Agricultural Mechanization embraces the use of tools, implements and machines for agricultural land development, crop production, harvesting, preparation for storage, storage, and on-farm processing. It includes three main power sources: human, animal, and mechanical. The manufacture, distribution, repair, maintenance, management and utilization of agricultural tools, implements and machines is covered under this discipline with regard as to how to supply mechanization inputs to the farmer in an efficient and effective manner.

Hand tool technology is the simplest and most basic level of agricultural mechanization: the use of tools and simple implements using human muscle as the main power source. Draught animal technology refers to implements and machines utilizing animal muscle as the main power source.

Mechanical power technology is the highest technology level in agricultural mechanization. It embraces all agricultural machinery which obtains its main power from other sources other than muscular power.

Within each of these three levels of mechanization technology, degrees of sophistication must be distinguished. For example, a simple locally made single-axle tractor without differential gears and gear box, a single axle tractor with gearbox and power-take-off, and a 70 kW tractor, are all mechanical power technology, but with a large difference in sophistication and capability.

Terminology such as "intermediate technology" and "selective mechanization" are either inappropriate or have no practical meaning. The term appropriate mechanization may be used, and refers to the level of mechanization and how it is used for a specific situation. "Appropriateness" can only be determined after carefully considering the technical, economic and social characteristics of each situation. No generalizations can be made concerning the appropriateness of a particular type of mechanization or particular agricultural tool, implement or machine for rural development. For the purpose of simplicity, the term mechanization in this Section will be used to cover all levels of mechanization technology and their degrees of sophistication.

INNOVATIONS IN AUTOMATED AGRICULTURAL MACHINERY AIMING FOR ULTRA-LABOR SAVING

In addition to KSAS, which assists farm operations by communicating with machinery and collecting and analyzing work progress, cultivation management and crop information, we are also focused on research and development that will further improve the efficiency of already mechanized work such as tilling and harvesting, in order to enable more precise work with minimized labor.

Kubota's roadmap of automation and unmanned operation has three steps. In 2017, Kubota launched its Agri Robo Tractor and in 2018, it launched Agri Robo Combine Harvester as Step 2 machines in Japan. In October 2020, Agri Robo Rice-Transplanter will also be available. Also, in Step 2, we are proceeding with further research and development such as sophistication of the control systems, advancing unmanned operation of periphery work, and further automation of field work.

In terms of Step 3, we plan to realize unmanned operation in multiple fields and remote monitoring of traveling on farm roads.

Further research and development are required, including the incorporation of automobile manufacturers' technology such as 3D dynamic maps, as well as further advancement of safety systems. It will also be necessary to establish high-speed communication infrastructure and mitigate the Road Traffic Act.

To further increase operational effectiveness of automated and unmanned agricultural machinery, also preparing for operation and management of multiple machines, supporting formation of optimized travel routes, and building a system that can collect, monitor and utilize the data from automated agricultural machinery.

FARM AUTOMATION TECHNOLOGIES

Farm automation will happen seamlessly, as we already have the technologies capable of bringing smart solutions to the table, such as automated greenhouse systems with profound analytics. There are three leading technologies that we can expect to see in the fields in the years to come: robots, autonomous machines, and drones. Each will represent another shift in the employment paradigm by altering the skills farmers require of employees, with GIS competence being the most significant.

The entire world is accepting GIS-powered precision farming with grace and gratitude, knowing that it will change and refine the way people produce, collect, distribute, and consume food.

Precision farming is surely going places. The benefits it offers are way too alluring for both companies and employees. And while you might think that these new benefits will set in motion huge job losses, you might be mistaken. After all, a little education never killed nobody.

ROBOTS & AUTONOMOUS MACHINES

Do you remember the 2005 classic I, Robot with Will Smith? Well, it might not be a classic, but there's no accounting for taste, right? Just like Will Smith's nemesis, robots in automated farming are here to substitute for people. In fact, robotic techniques will substitute for seeding, weeding, spraying, picking, and many other tasks and routines, which leaves humans with one single task: to learn how to service the robots.

Oh, and one more thing: don't forget that robots are not necessarily two-footed humanlike creatures. They can come in the form of tractors, combines, and other types of machines. Automated farming robots and drones are there to deal with four major missions.

Autonomous precision seeding

Seeding is where it all begins in agriculture, right? You can start to forget about walking an extra mile (literally) in the fields, as automated drone seeders have you covered. They are efficient and fast; ten drones can plant up to 400,000 trees a day. Hence, while drones are the major agents of crop quality and safety monitoring, autonomous precision seeding technology lies at the heart of change for those working in the agricultural sector of the future.

IS-powered weeding

As you might have guessed, farming automation boils down to cutting down on the number of people out in the fields. GIS-powered precision weeding, powered by IoT robotics, is a way to ensure the crop makes it to harvest safe and sound.

Precision picking

All plants and crops have their own features when it comes to ripening. Some do it faster; some do it slower. But they all need to be picked one day or another. How can we know when the day comes and how can we know in which part of the field to harvest? Another thing is that many berries, fruits, and vegetables are quite tender and fragile, often suffering from human handling. Hence, precision picking is now being used as one of the main tools for accelerating and increasing the efficiency of the picking process.

Non-traditional farming robotics

Have you ever heard of indoor growing, or so-called vertical farms? You may end up eating salad tonight that's been seeded, grown, and picked in a vertical farm. Smart vertical farm lighting and monitoring solutions are transforming agriculture, as vertical farms can be fully autonomous. Indeed, it might take just five to ten employees to take care of an entire farm. For example, the Iron Ox company boasts of running a fully automated 8,000-square-foot farm. Shields of hydroponic trays and cloud-controlled robots take care of the leafy greens that seem to be ubiquitous. And yes, smart LED lights substitute for the sun, while such farms use 90% less water than their conventional counterparts.

HOW FARM AUTOMATION WILL LEAD TO DEMAND FOR NEW SKILLS

Now that we know that agricultural automation will come in four major modes — seeding, weeding, picking, and non-traditional farming — let's define the skills that those wanting to continue their agricultural employment will have to acquire in the long term. First and foremost, you must understand that the question is not how to make an automatic farm. The situation the industry finds itself in focuses on integrating automation in agriculture into manual labor to achieve the best result possible. We are still far from omnipresent full-fledged automatic farm systems. Agricultural professionals must acquire five basic skills:

Common software skills

Farm workers must be able to turn on a tablet, connect to a Wi-Fi network, send messages, compress files, etc., as well as be comfortable handling other electronic devices.

Microsoft Office / G Suite

These are some of the major team communication and data sharing tools. Workers need to know how to open a spreadsheet, insert data, and extract data.

SPSS or SAS

Modern farming has a lot in common with statistics. Therefore, workers will have to know how to deal with data analytics tools.

Geographic Information System (GIS)

The holy grail of precision farming, GIS is something workers must at least have some familiarity with. GIS-based agriculture is the core of effective field operations and planning.

IT aptitude

This is not a skill but rather the ability to acquire all the skills needed to become a valued employee in the new agricultural industry. An aptitude to acquire IT skills, which is basically the ability to learn, will let workers acquire the other four skills listed above and every other skill that their employer might eventually require.

Robots are not here to cause a farming condition overload; they are here to help people become more advanced, cleverer, and more productive while being physically idle. The Association for Advancing Automation calls farm workers of the near future fleet managers taking care of robotic farming systems and autonomous vehicles.

Modern workers will develop new equipment functions, take control of equipment movement, or simply turn equipment on and off (oh, those old but gold IT support days). The principle at work is simple: the more functions devices will have, the more attention they will need from humans. Also, devices will require maintenance. The more devices there are, the more service personnel farms will need.

It's not about removing people at all; it's about changing the scope of responsibilities and boosting safety. What's more, robots are only executives. Agricultural science and complex decisions will still fall on humans' shoulders. All those machines are to be built, sold, and repaired before they oust humans, right? Finally, all the data collected by robots and drones will require human analytical skills. The machines are here only to refine automation in agriculture by letting people make accurate decisions on land erosion, crop health, water runoff, and many other aspects.

FARMING AUTOMATION SOLUTIONS

Before we move to the discussion of solutions that new agricultural automation technologies offer, let's come to terms with one thing: farming digitalization is not to blame for the shortage of jobs in the agricultural sector. On the contrary, it's one of the major agents of positive change, advocating for more rights and safety for agribusiness employees. If you're still wondering whether change is due, let's have a look at the number of active IoT connections in agriculture across Europe.

How will farm automation make the lives of those in the agricultural business easier, regardless of their position? There are quite a few places where robotics might come in handy when transforming agriculture.

Monitoring and analytics

Aside from the holy trinity of seeding, weeding, and picking, agricultural software and hardware is being heavily used for monitoring and analyzing crops, which is essential for preserving them until harvesting. It's no wonder why the global market for agricultural robots is expected to grow from \$4.9 billion in 2021 to \$11.9 billion by 2026 according to Markets and Markets. This is a CAGR of 19.3%. Agricultural robotics help farmers detect maladies, eradicate them, and live up to market expectations by preserving crops from tillage to harvesting.

Guidance

Timely and refined decision-making is what successful farming boils down to in 90% of cases. The other 10% is luck due to fluctuating weather conditions. Meanwhile, all other factors such as soil quality, crop ripeness, and crop health can easily be defined with the help of agricultural robotics. Yet it seems like the technology has already gone extra. In accordance with Riaz (2021), agricultural robotics have made a step forward by enabling multi-criteria decision-making (MCDM), which lets farmers not only define what's wrong and right but also rank alternatives from low to high priority.

Irrigation

Robotic irrigation cuts down farming costs. As a matter of fact, we won't talk too much about the value of robotic irrigation and the principles of its work. Let us show you something really fascinating instead. Here is a photograph of croplands in Ogallala, Nebraska, in the United States. Can you guess which of these circles have been irrigated automatically and which enjoyed manual irrigation?

It looks like there actually is something special about farmers who wonder how to make an automatic farm. Automation in farming is all about protecting people from exploitation and teaching people new things that will, at the very least, grant them unique knowledge and experience required for further employment.

Change is inevitable

Farm automation is not something that came out of the blue and started forcing agribusinesses to adopt it using sticks rather than carrots. It has resulted from a long-lived process of introducing digital technologies into the agricultural industry, which has eventually proven that farmers are better off with automated solutions than without.

As we enter the third decade of the 21st century, farming automation solutions are only going to double their presence on farms, simultaneously changing the role people play in the farming process. Yet there's no need to worry. No job loss is intended, as farming automation is rather about enjoying the process and being sure of the result your fields provide and not about farming condition overload. Yes, jobs will change their essence, but they will still be there: safe, legal, and advanced.

HYBRIDIZED PLANTS AND HARVESTING

When designing and running automation equipment, uniformity is king. If the material being handled is all the same size, shape, weight, texture, firmness, it's much easier to pick, move, separate, inspect, etc. This remains true in agriculture; and dealing with living, growing biological shapes presents interesting challenges. With increasing automation taking place in agriculture, it makes sense that farmers might again look to selective breeding and GMO crops to solve some of these challenges in the modern age. But now, instead of just optimizing for raw yield, a farmer might choose a variety of crop that is known to grow more uniformly with fruit of a certain size; or stalks of a particular height that can more easily be harvested by a tractor with a mechanical picker; or maybe a variety that all comes ripe at the same time so a whole field can be harvested at one time. So in this modern age, many more parameters need to be optimized for than just yield or taste.

A blueberry that tastes amazing, but is very delicate and must be hand-picked, will not be as desirable to a farmer as a blueberry with a tougher skin that can be machine-harvested and survive through the conveyor belts of the washing and packaging process. And in fact the price the end customer pays will directly reflect the ease of harvesting, packing, transport logistics, and spoilage. Of course, the intent is to maximize all of these parameters simultaneously, including taste and quality; but food engineering, as will all engineering, is truly the art of managing tradeoffs. For this reason, oddly-shaped heirloom tomatoes that must be hand-picked and hand-packed will always be more expensive.

If aspects of the plant can be made more homogenous, then the machine to handle that crop can be simpler, which often means less expensive and more reliable. But even if the plant cannot be made more homogenous, there are still ways to automate its handling; it's just that the system complexity may have to scale up to handle the irregularities. More sensors can be added, or more degrees of motion can be built into the actuators. Machines can be built to handle even the most odd-ball of shapes.

But this doesn't come free, the complexity of such a machine often goes up by orders of magnitude, and with it the cost to purchase and maintain. In agriculture, where volume is high and prices are low, it's very important to keep equipment costs down and robustness/reliability up. The equipment has to work or it won't get used. In farming and food production there is often a narrow window of time to harvest, process, and deliver the product before it goes bad. So a large-scale farmer is unlikely to have time to futz around with a finicky piece of equipment. For this reason, the more reliable tool will generally win out in the field and in the packing house. Human hands and eyes are very easy to train, and then very reliable for short timespans of a couple of hours. Cameras and end effectors (technical name for grippers) are much harder to train to do the job right.

So ideally the crop itself would not require complex end effectors to be developed to handle it. Better if the crop is known to be uniform enough that simple, robust metal rails, sieves, combs, or some other simple mechanism can be used to perform a function over and over and over again with no babysitting and no maintenance. For this reason, crop science, genetics, informatics, and machine automation all go hand-in-glove to streamline this process. So I believe that the advances in food production that we will see over the coming decade will be due in equal part from advances in machine automation as well as from efforts made in the life-sciences to engineer better plant hybrids.

The benefits of using hybridized plants coupled with customized machinery have already been brought to bear on certain crops such as blueberries, grapes, strawberries and tomatoes.

Mechanical grape harvesting has become the norm in California, where 80% of wine grapes are harvested by machine. One machine can harvest 15 to 20 tons of grapes per hour, work that would normally take 30 or more human pickers, at less than half the cost of picking by hand.

Labor costs in the domestic strawberry industry are roughly $1b per year and it is becoming harder to find people to perform the backbreaking task of picking the berries from the knee-height bushes. Reducing the number of hands that touch strawberries on their way to the end consumer has cost and safety benefits also, while alleviating the problem of finding people to do the work. Strawberry picking robots can pick 8 acres of strawberry plants in one day and can replace the work of 30 human pickers. Harvest Croo Robotics and Agrobot are two startups competing to serve this expanding market.

DEVELOPMENTS IN MECHANISATION TECHNOLOGY

Mechanization of agriculture is an essential input in modern agriculture. It enhances productivity, besides reducing human drudgery and cost of cultivation. Mechanization also helps in improving utilization efficiency of other inputs , safety and comfort of the agricultural worker , improvements in the quality and value addition of the produce. Efficient machinery helps in increasing production and productivity, besides enabling the farmers to raise a second crop or multi crop making the Indian agriculture attractive and a way of life by becoming commercial instead of subsistence. Increased production will require more use of agricultural inputs and protection of crops from various stresses. This will call for greater engineering inputs which will require developments and introduction of high capacity, precision, reliable and energy efficient equipment.

Looking at the pattern of land holding in India, it may be noted that about 84% of the holdings are below 1 ha. There is a need for special efforts in farm mechanization for these categories of farmers to enhance production and productivity of agriculture.

In the existing scenario of land fragmentation and resulting continued shrinkage of average size of operational holdings, percentage of marginal, small and semi-medium operational holdings is likely to increase. Such small holding makes individual ownership of agricultural machinery uneconomic and operationally unviable. 'Custom Hiring Centers of Agricultural Machineries' operated by Cooperative Societies, Self Help Groups and private/rural entrepreneur are the best alternative in enabling easy availability of farm machineries to the farmers and bringing about improvement of farm productivity for the benefits of Small & Marginal farmers.

IMPORTANCE OF TECHNOLOGY

Technology in agriculture affects many areas of agriculture, such as fertilizers, pesticides, seed technology, etc. Biotechnology and genetic engineering have resulted in pest resistance and increased crop yields. Mechanization has led to efficient tilling, harvesting, and a reduction in manual labor. Irrigation methods and transportation systems have improved, processing machinery has reduced wastage etc., and the effect is visible in all areas.

New-age technologies focus on robotics, precision agriculture, artificial intelligence, block chain technology, and more. Some technological advancements that have innovated agriculture:

- Improved productivity from mechanization of agriculture. To reduce manual labor and make the processes faster, combine harvesters are finding greater use. Indian farming is characterized by small landholdings, and the need is to partner with others to take advantage of modern machines.
- Climate/ weather prediction through artificial intelligence. A major advance in agriculture is the use of artificial intelligence (AI). Modern equipment and tools based on AI enable data gathering and assist in precision farming and informed decision-making.

Drones, remote sensors, and satellites gather 24/7 data on weather patterns in and around the fields, providing farmers with vital information on temperature, rainfall, soil, humidity, etc.,

However, AI finds slow acceptance in a country like India where marginal farming, fragmented landholdings, and other reasons act as impediments. But there is no doubt that technologies based on AI can bring precision to large-scale farming and lead to an exponential rise in productivity.

- Resilient crops developed via use of biotechnology. Agriculture refers to a wide resource of methodologies that include traditional breeding methods, genetic engineering, and development of microorganisms for agriculture. Generally speaking, genetic engineering uses the understanding of DNA to identify and work with genes to increase crop resistance to pests, and the development of high-yielding varieties also make improvements to livestock.
- The spinoff of biotechnology in agriculture has resulted in all-around benefits for farmers and end consumers. Though some controversial approaches have led to resistance in the adoption of biotechnology, there is no doubt that the future of agriculture is heavily dependent on SAFE biotechnology, given the changing climate and increase in population.
- Improving farm yields and supply chain management use Big Data. The collection and compilation of data and its further processing to make it useful for decision-

making/problem-solving are expanding the way big data functions. Big data is slated to play a major role in smart farming, and the benefits percolate across the entire supply chain and the markets. Agriculture is becoming larger, and it depends on a large number of variables.

This is resulting in greater collection and use of complex data, which has to be meaningfully interpreted and managed. Data can be from external sources such as social media, supplier network, markets, or from sensor/machine data from the fields. Transformation of agriculture from using big data is taking place that affect crop yield, supply chain management, yield prediction, etc.

THE NEED FOR MECHANIZATION / PRODUCTIVITY ENHANCING TECHNOLOGY.

Farm Power — consisting of manual labor, agricultural tools, draught animals, tractors, implements, equipment, and machinery — is an essential farm input. In almost any agricultural production system the annual expenditure on farm power, whether on labor, draft animals, or fuel and depreciation of machines, largely exceeds the costs of other inputs such as agro-chemicals and seeds. In many developing countries, agricultural production and food security are adversely affected because of insufficient use of farm power, low labor productivity and/or labor scarcity. The need to improve agricultural labor productivity is increasingly recognized. In the case such as pump sets for irrigation, the need for machinery is undisputed. Rather than agricultural mechanization, it would be preferable to use the term Farm Power or Labor Productivity Enhancing Technology, to recognize not only the importance of manual labor and handtools, draft animals, and mechanical power, but also other issues related to labor scarcity, such as cropping- and farming systems.

Finding solutions to environmental problems in agriculture requires (improved) agricultural tools and machinery, for example for soil tillage and pesticide application, the latter also addressing health concerns. Similarly, machines are required to assist with post-harvest loss reduction and on-farm processing. Thus it is now (again) recognized that agricultural mechanization is crucial in the fight against hunger and poverty, and at the same time to address environmental and health concerns. In order to avoid recurrence of the past mistakes such as described in the introduction, formulation of efficient mechanization strategies are required.

The term mechanization is unfortunately often very narrowly perceived while its real purpose, namely, enhancing productivity of land and labor is often not well understood. In fact an agricultural mechanization strategy ought to be part of an agricultural technology strategy, which is to be part of an overall agricultural development strategy. In this context, three principal purposes of mechanization may be summarized as follows:

- Increase in labor productivity. The introduction of machinery to substitute for labor ("laborsaving") is a common phenomenon associated with the release of labor for employment in other sectors of the economy or to facilitate cultivation of a larger area with the same labor force.

- Increase in land productivity. The purpose of mechanization is here to produce more from the existing land. Machinery is a complementary input, required to achieve higher land productivity, for example, through the introduction of pump sets, or faster turn-around-times to achieve higher cropping intensity. In labor surplus economies, net labor displacement or replacement should be avoided.
- Decrease in cost of Production. Introduction of a machine may lower production costs or offset increased costs of draft animals or labor.

Usually, in various degrees, a combination of the three objectives will be achieved. Additional benefits to the user may be associated with a reduction in the drudgery of farm work, greater leisure, or reduction of risk. These are subjective benefits and difficult to translate into cash. Frequently mechanization increases an individual's workload, can be hazardous to health and may reduce the social interactions associated with farm work.

THE ADOPTION PROCESS FOR MECHANIZATION/LABOR PRODUCTIVITY ENHANCING TECHNOLOGY

When reviewing the process of applying labor-saving (or labor productivity enhancing) innovations in agriculture, it is a serious but frequently made mistake to assume that this can be achieved only through applying mechanical engineering technology. In this context, nine different stages in the process of enhancing labor productivity may be distinguished.

- **Stage I:** Application of improved Hand tool Technology. This process started in prehistoric times when early civilizations developed stick and stone tools which were the only means to enhance labor productivity. In many parts of the world, handtools are the only technology used in agriculture, and even in highly mechanized agricultural systems, improved handtools are still important.
- **Stage II:** Draft animal power application. At this stage animal muscle power is substituted for human power, a process which already started in ancient civilizations. A large variety of implements and machines have been developed which use animals as the principle power source.
- **Stage III:** Stationary Power Substitution. Mechanical power is substituted for human and animal power, used in stationary operations. Stationary operations are mechanized first because motive power sources required to move across the field are technically more complex and therefore require higher investment. Typically, operations mechanized at this stage are paddy dehusking, grain milling, pumping water, and threshing.
- **Stage IV:** Motive Power Substitution. At this stage, substitution of mechanical power for muscle power takes place for field operations. It focuses on power-

intensive field operations (for example, plowing), and machinery is of relatively simple design, and easy to operate. Mechanization is still straightforward, and crop production practices are usually unchanged. At Stage III and IV, mechanization takes advantage of lower costs of new power sources as compared with traditional ones.

- **Stage V:** Human Control Substitution. At this stage the emphasis is on substitution of the human control functions. Depending on the complexity of the control function and the degree of its mechanization, machinery becomes increasingly complicated and costly. A potato lifter is simple in design, but fruit and cotton harvesting machinery are complex and expensive.
- **Stage VI:** Adaptation of Cropping Practices. This stage features the adaptation of the cropping system to the machine. For example, removing weeds in broadcast crops cannot be done with machines but row seeding and seed drills may be introduced to facilitate mechanization of weeding. Other examples include the increase in row distance to accommodate heavier and larger machinery to speed up field operations.
- **Stage VII:** Farming System Adaptation. The farming system and production environment is changed to facilitate further increase in labor productivity and to benefit from economies of scale, necessary to make the investment in expensive machinery financially feasible. An example of this is the rapid decline of mixed farming systems in Europe since the late 1960s when farmers specialized in either dairy, poultry, hog, or crop production. Some crops which are difficult to mechanize may disappear if acceptable substitutes become available, or if these can be produced in countries with low labor costs. At this stage, investments in land development, land consolidation, and rural infrastructure are often needed to facilitate advanced degrees of mechanization.
- **Stage VIII:** Plant Adaptation. This stage features the adaptation of the plant and animal to the mechanization system. Mechanization has advanced to a stage were engineering alone can no longer provide further gains in labor productivity. Breeders increasingly take into account the suitability of new varieties for mechanized production.
- **Stage IX:** Automation of Agricultural Production. This stage is progressing in countries with high labor costs and sophisticated demands on production and quality. Examples are automated rationing of concentrate feeding for individual dairy cows based on their milk production, and sprinkler irrigation systems activated by soil moisture.

AGRICULTURAL MACHINERY AND TECHNOLOGY & THEIR USAGE IN AGRICULTURE

Various types of modern agriculture machinery and technology are used in different agricultural operations nowadays. Different levels in crop production include – Primary and secondary level tillage of the soil, Seeding and planting, Cultivation, Fertilizer application and distribution, Pest control, Harvesting, Irrigation, Drainage,Transportation, Storage, Handling the residues of earlier crops etc.

Since the olden times, animals were the primary source of energy when it was about tough jobs in farming. Later on, steam power started replacing the same. And then gas-powered tractors took over, followed by diesel engines. In the developed nations, it caused lowering the number of farmworkers, however, farm production continuously increased with the use of agriculture machinery.

Technology in agriculture has transformed and increased production and quality of produces. In modern times, farmers who are doing heavy works on farms using traditional and old agriculture tools are wasting their health and time. A tractor that used to be known as a technological genius in the agricultural field is old news now. The modern farm machinery has upgraded the agricultural industry for the best. Some of the essential and most used machinery are Combine or Combine Harvester, Rotavator or Rotary Tiller, Plough or Plow, Tractor Trailer, Power Harrow, Leveler, water bowser, ripper machine, and disc harrow. Below we have mentioned a few of the latest agriculture machinery and their uses in farming.

Plough

Plough, a primary tillage equipment works as a tractor implement attached with a tractor and helps in tilling the land efficiently. This remarkable piece of agricultural machinery is used for preparing the land in the planting season. Either its about turning an all new land into a farm or just preparing existing farm to sow seeds, various types of plough (mould-board plough, reversible plough, disc plough) allow farmers to turn the soil into a nutrient rich seedbed for better plant growth.

Fig. 1: PloughPlough

Combine Harvester

Bigger in size a Com-bine works like a comb cutter for cutting the mature crops with threshing at the same time. A combine is one of the most upgraded machinery that helps farmers by saving long hours of the day wasting in harvesting activities.

Fig. 2: Combine Harvester

The multi-crop combine harvester is designed in such a way that it is compatible for harvesting multiple grain crops such as wheat, corn, barley, grain sorghum, soybeans, oats, sunflower, rice. A modern multi-crop harvester makes harvesting faster so that the fields can be open for the next crop in minimum period of time, and the probability of production increases within the period.

Rotavator or Rotary Tiller

A rotavator or rotary tiller is a versatile tillage equipment that uses a series of blades, to turn up the soil. Rotary tillers are quite popular among farmers to make easy the preparation of seedbed in the farms. The equipment is mainly used for breaking soil in lawns, farms, gardens, etc. The use of rotary tillers has significantly increased in agricultural operations because of its simple built and high-end efficiency as a tillage implement.

Fig. 3: Rotavator or Rotary Tiller

Disc Harrow

The disc harrow is one of the essential tillage implements that farmers can invest into, to increase their production. This tillage equipment is also used in secondary tillage process as it helps in breaking up lumps of the soil easily, efficiently and faster to give it a rich finish when it comes to prepare a final seedbed. It also allow farmers to control weeds around the plants once the crop is planted and grown.

Fig. 4: Disc Harrow

Leveler

This tractor implements make the soil more levelled, smooth and surface obtained, enabling it to create a moist environment for crops and reducing consumption of fertilizers, seeds, chemical and fuel.

Non inverter peanuts digger shaker

Based on the conclusions drawn from field observation study, a tractor mounted new digger shaker was designed (Fig. 5) and constructed (Fig. 3). The tractor rides on a growing bad and pulls the digger. The digger continually uproot the plants, gently raise them while shaking them to clean from dirt and prepare a non invert, uniform and organized windrow. Two rotating square rod bars (1) dig and penetrate the ground beneath the plant, a first conveyor (2) mounted in moderate inclination with strong round rod bars, penetrate the ground too, gently raise the soil and plants and convey them to a double inclined square hollow structure bars conveyor (3) that shake them free of soil and convey the uprooted plants back to the ground where the peanuts are placed on the ground while the foliage cover them. Two closed vertical cylinders (4) are mounted behind the digger on two parallelograms with hydraulic reduced weight system, are rotated from ground, which reduces the width of the windrow to a size that is suitable for the pickup harvester. In front of the digger there is a depth control mechanism (5) that continually keeps the desired uprooting depth in order to prevent peanuts losses.

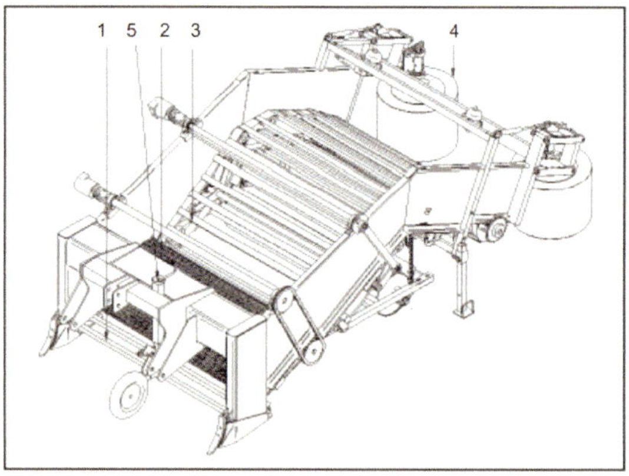

Fig. 5: Isometric view of new non inverter peanuts digger

Fig. 6: New non inverter peanuts digger in field

13

ADVANCES IN MECHANIZATION AND AGRICULTURAL AUTOMATION

The mechanization of farming practices throughout the world has revolutionized food production, enabling it to maintain pace with population growth except in some less-developed countries, most notably in Africa. Agricultural mechanization has involved the partial or full replacement of human energy and animal-powered equipment (e.g. plows, seeders and harvesters) by engine-driven equipment. Most of this is tractor driven and to a lesser extent self-propelled equipment (including harvesters, sprayers, fertilizer applicators, planters and seeders). Agricultural mechanization has been pioneered in North America and Europe and more recently in Japan, and is now spreading rapidly throughout the world.

Notwithstanding such progress, a significant element of human and animal powered mechanization remains, particularly in the poorer regions of the world. The importance of enhancing and upgrading such mechanization practices prior to the almost inevitable transition to engine-driven equipment is now well recognized. Automation of agricultural mechanization is an intensive area of research and development with emphasis on enhancement of food quality, preservation of operator comfort and safety, precision application of agrochemicals, energy conservation and environmental control. Automation applications will be orientated towards and assist in the attainment of environmentally friendly and sustainable systems of agricultural and food production.

However, the difficulties in matching environmental concerns and sustainability with an ever-increasing world population cannot be underestimated especially in the developing countries. Thus, there may be a tension between maximizing food production on the one hand and implementing sustainable development and environmental protection systems (e.g. erosion control) especially, in poorer regions, where the demand for increased food production follows logically from an increasing population.

MECHANIZATION IN ALLIED SECTORS

In the dairy and livestock sector, which is largely integrated with an associated industry, the adoption of mechanization has been more common. Milking machines, fodder handling and feeding machines, harvesting systems in abattoirs, etc. are examples. In the fisheries sector the adoption of mechanization has been even more, and it has reaped major benefits. Examples here, range from mechanised fishing boats which deploy propulsion and motorised net handling equipment, fish pond management equipment, etc. Factory ships in the fisheries sub-sector are probably the largest format of machines and mechanization deployed in the agricultural sector.

On farm fields, mechanised tilling and harvesting are now most commonly used and India is the largest market for tractors and a variety of harvester machines. Irrigation pumps, fertigation systems and pesticide applicators are other mechanised forms of farm inputs. However, given the scale of India's agricultural sub-sectors, mechanization is yet to penetrate all across regions and farm types. One of the main limitations to deploying farming machines, is the capacity of a farm to efficiently own such equipment. Small farmers will continue to be the mainstay of Indian agriculture and the concept of custom hiring centres and mechanised services is therefore most apt in this situation.

Mechanization is one factor that has had a significant effect on TFP since the beginning of modern agriculture. Mechanized harvesting, for example, was a key factor in increasing cotton production in the last century. In the future, mechanization will also have to contribute to better management of inputs, which will be critical to increasing TFP in global production systems that vary widely among crop types and regional economic status.

For example, a scarce, basic resource that will have to be managed much better is water, a critical input in agricultural production. Both the efficiency and effectiveness of water use will have to improve dramatically.

Today, approximately 70 percent of withdrawals of fresh water are used for agriculture. By 2025, 1.8 billion people are expected to be living in areas with absolute water scarcity, and two-thirds of the world population will live in water-stressed areas. Improving water management will have to be achieved by more efficient irrigation technology and higher efficiencies in whatever technologies farmers are currently using.

THE IMPACT OF MECHANIZATION ON PRODUCTIVITY

Agricultural mechanization, one of the great achievements of the 20th century, was enabled by technologies that created value in agricultural production practices through the more efficient use of labor, the timeliness of operations, and more efficient input management with a focus on sustainable, high-productivity systems. Historically, affordable machinery, which increased capability and standardization and measurably improved productivity, was a key enabler of agricultural mechanization.

In the 19th century, as our society matured, a great many innovations transformed the face of American agriculture. Taking advantage of a large labor base and draft animals, farmers had been able to manage reasonable areas of land. This form of agriculture was still practiced in some places until the middle of the 20th century.

Early innovations were implements and tools that increased the productivity of draft animals and assisted farmers in preparing land for cultivation, planting and seeding, and managing and harvesting crops. The origins of the John Deere Company, for example, were based on the steel-surfaced plow developed by its founder. This important innovation increased the productivity of farmers working in the sticky soils of the Midwest.

A major turning point occurred when tractors began to replace draft animals in the early decades of the 20th century. Tractors leveraged a growing oil economy to significantly accelerate agricultural productivity and output. Early harvesting methods had required separate process operations for different implements. With tractors, the number of necessary passes in a field for specific implements was reduced, and eventually, those implements were combined through innovation into the "combination" or combine harvester.

For most of the 20th century, four key factors influenced increases in the rate of crop production: more efficient use of labor; the timeliness of operations; more efficient use of inputs; and more sustainable productions systems (Table 1). These four drivers played out at different rates in different crop production systems, but always led to more efficient systems with lower input costs. Technological innovations generally increased mechanization by integrating functional processes in a machine or crop production system and by making it possible for a farmer to manage increasingly large areas of land.

By the late 20th century, electronically controlled hydraulics and power systems were the enabling technologies for improving machine performance and productivity. With an electronically addressable machine architecture, coupled with public access to global navigation satellite system (GNSS) technology in the mid-1990s, mechanization in the last 20 years has been focused on leveraging information, automation, and communication to advance ongoing trends in the precision control of agricultural production systems.

In general, advances in machine system automation have increased productivity, increased convenience, and reduced skilled labor requirements for complex tasks. Moreover, benefits have been achieved in an economical way and increased overall TFP.

FROM MECHANIZATION TO CYBER-PHYSICAL SYSTEMS

Today's increasingly automated agricultural production systems depend on the collection, transfer, and management of information by ICT to drive increased productivity. What was once a highly mechanical system is becoming a dynamic cyber-physical system (CPS) that combines the cyber, or digital, domain with the physical domain. The examples of

CPS reviewed below suggest the future potential of ICT for achieving the target TFP of 1.75 and beyond.

Precision Agriculture

Precision agriculture, or precision farming, is a systems approach for site-specific management of crop production systems. The foundation of precision farming rests on geospatial data techniques for improving the management of inputs and documenting production outputs.

As the size of farm implements and machines increased, farmers were able to manage larger land areas. At first, these large machines typically used the same control levels across the width of the implement, even though this was not always best for specific portions of the landscape that might have different spatial and other characteristics.

A key technology enabler for precision farming resulted from the public availability of GNSS, a technology that emerged in the mid-1990s. GNSS provided meter, and eventually decimeter, accuracy for mapping yields and moisture content. A number of ICT approaches were enabled by precision agriculture, but generally, its success is attributable to the design of machinery with the capacity for variable-rate applications. Examples include precision planters, sprayers, fertilizer applicators, and tillage instruments.

The predominant control strategies for these systems are based on management maps developed by farmers and their crop consultants. Typically, mapping is done using a geographic information system (GIS), based on characteristics of crops, landscape, and prior harvest operations.

Sources of data for site-specific maps can be satellite imaging, aerial remote sensing, GIS mapping, field mapping, and derivatives of these technologies. Some novel concepts being explored suggest that management strategies can be derived from a combination of geospatial terrain characteristics and sensed information. All of these systems are enabled by ICT.

A competitive technology for map-based precision farming is on-the-go sensing systems, based on the concept of machine-based sensing of agronomic properties (plant health, soil properties, presence of disease or weeds, etc). The immediate use of these data drives control systems for variable-rate applications. These sensor capabilities essentially turn the agricultural vehicle into a mobile recording system of crop attributes measured across the landscape. In fact, current production platforms are increasingly becoming tools for value-added applications through ICT.

Precision Guidance

Around the turn of the 21st century, GNSS technology had become so precise and accurate that it had outpaced the requirement for the early phases of precision farming and become commercially viable for enabling a number of automatic-guidance applications (Han et al., 2004). Advances in GNSS technologies include decimeter to centimeter accuracy by using signals from a geospatially known reference point to correct satellite signals. One premium example is a real-time kine-matic global positioning system (RTK-GPS) technology (Figure 3a) that reduces fatigue and lowers the skill level required to achieve high-performance accuracy in field operations.

FiIn short, in less than 20 years, GPS technology went from being an emergent technology to a robust, mature technology that has optimal capabilities for production agriculture. A number of solutions are emerging today (Figure 3a) for achieving high-precision accuracy through various reference-signal configurations (e.g., RTK-GPS, multiple satellite systems, sensor fusion with complementary sensors, and multiple sources of corrections).

Operator-guidance aids that provide feedback to the operator about required steering corrections through audio and visual cues were the first systems on the market for precision guidance. This feature allowed a vehicle system to follow paths parallel to prior operations across a field. These types of systems worked well at decimeter accuracy and required no major control-system integration into the vehicle.

The major benefits of these systems were to reduce overlap/underlap in field operations with extremely wide implements, typically for spraying chemicals and fertilizers. The decrease in overlap meant the parsimonious use of resources. The decrease in underlap meant that chemicals and fertilizers were applied to every part of the field.

On the next level of evolution, automatic guidance systems appeared that managed steering for an operator through automatic control. Automatic guidance systems enabled precision operations depending on the type of GNSS signal and how it was integrated into the requirements of the agricultural operations.

GNSS technology enabled the management of inputs such as seed, pesticides, and fertilizers with precision across the field. For example, the chemical application to buffer zones and grassy waterways was reduced based on sensing of the field location of these features. John Deere's software product, SwathControl Pro (Figure 3b), enabled farmers to manage the definition and execution of this capability.

GNSS technology provided the reference signal that enabled accurate vehicle location at the GNSS sensor, but precision control of the machine required several additions to the system (e.g., attitude correction, inertial sensors, implement control). With these features, a mobile CPS could correct the attitude of the vehicle on uneven terrain and manage the vehicle system path for precision in the execution of complex functions.

The ultimate in un-manned automation is the capability of driving complete field patterns under autonomous management of the tractor-implement functions without frequent operator intervention. Figure 3c shows one commercial example of the execution of this concept. The figure shows a very rudimentary form of path planning, integrated with automatic guidance, that can increase productivity by managing the paths a vehicle must follow. Path management can be programmed to reduce time loss caused by navigation (e.g., turning around) and implement management.

Like precision agriculture, precision guidance creates data from its precision operations that could be used in crop management. Examples of these data include information on the "as-applied" state of operations, vehicle paths, and operational state variables. The data can then be used to meet the needs of other ICT in systems automation and optimization.

System Automation and Control

Until recently, automation has been focused on functions that depend on GNSS or direct sensing. However, processes that lend themselves to control based on the attributes of soil and crop properties are also being investigated. Some initial applications of these, which were coupled with GPS, mapped the yield and moisture of harvested crop operations.

It is also possible to use sensing of soil or crop properties—such as controlling the cut-length of a self-propelled forage harvester (SPFH)—as part of a combination of techniques to increase machine system productivity. In this example, the cut-length is the section length into which a tree, or forage plant, is cut. When an SPFH is operated with static cutting settings, independent of the size of the forage plant, it can consume a significant amount of energy in cutting forage for ensiling (storage in silos).

HarvestLab™, a sensing technology, uses near infrared (NIR) reflectance sensing to detect the moisture content of forage and adjust the cut-length of harvested material. This control strategy can significantly reduce the energy consumption for harvesting forage with no degradation in the ensiling process. The results are a significant reduction in fuel consumption in the harvest operation and a high-quality cut, which enables proper forage preservation.

NIR sensing has often been used in the laboratory and in grain processing and storage to measure properties (e.g., moisture oil and protein content) of biological materials, which contributes to value-added uses of corn, cereal grains, and forage. As these technologies mature, ICT has the potential to connect information about constituent properties to downstream processes.

Machine Communications

The automation methods described above generate massive amounts of data. However, the data are not limited to on-vehicle storage or even to on-the-go decision making. Inter-machine communication greatly increases the potential of these systems.

In the last few years, the commercial application of telematics devices on machines has been increasing in agriculture, thus empowering a closer connection between farmers and dealers in managing machine uptime and maintenance services. Other applications for machine communication systems include fleet and asset management.

In addition, inter-machine communications are expanding machine system data applications, such as diagnosing and prognosticating machine health. Inter-machine communications can also include implements and tools (e.g., monitoring seeding rate in tractor implement applications). Functionally, a modern, high-end agricultural machine system is effectively a mobile, geospatial data-collection platform with the capacity to receive, use, sense, store, and transmit data as an integral part of its operational performance.

As we strive for higher TFP levels, these high-end applications are moving toward systems with increasingly advanced ICT capabilities, including data communication management from machine to off-machine data stores. Other ICT capabilities under development include vehicle-to-vehicle operations management in the field.

It is clearly within the vision of the industry to develop advanced capabilities (such as those listed below) that leverage these ICT innovations:

- machine knowledge centers that enable improved design, faster problem resolution, and higher system productivity, increased uptime, and lower operating costs
- tores of agronomic knowledge that can lead to optimization of farm-site production systems
- stores of social knowledge related to customer or consumer value-drivers

As ICT continues to penetrate production systems, a massive network is being developed of machine systems that are platforms for value creation—well beyond productivity from agricultural mechanization intended for the farmer or the farm site. These systems are collecting and managing information with potential value in downstream value-chain operations that use crop or drive systems to achieve environmental sustainability.

Worksite and Value Chain Productivity

The next step in automation and control is to move beyond individual vehicle systems to the optimization of production systems and farm worksites. To achieve this goal, we have developed the beginnings of vehicle and machine systems that can both sense and control with precision. These systems can be driven by data from a variety of sources to provide precision control. For example, they are capable of collecting, storing, and transferring information about the crop, field, and machine state at the time of field operation. They can also receive data from public and private data sources.

Furthermore, data collected by machines can be transferred to farm-management systems as well as to public and private sources that require information about production management for quality, compliance, or value-added purposes. Thus, we are entering an era of emerging field and farm optimization systems that can drive up TFP of the worksite, including machines, geographies, and cropping systems.

As intelligent mobile equipment for worksite solutions has evolved over the last 20 years, agricultural mechanization has also evolved from a bottom-up integration of the foundations of ICT applied to basic mechanization systems required for crop production. The primary machine capabilities of precision sensing, advanced control systems, and communications have created the potential for the emergence of CPS from production agricultural systems.

Although these advanced technologies are not uniformly distributed among platforms and production systems, where they exist, there are opportunities to leverage ICT to increase production systems capabilities. Looking ahead, it is expected that the business value of ICT will expand to additional platforms.

Technologies integrated on vehicles must work seamlessly with other systems. Drawbacks of some initial attempts for ICT capabilities have been the significant time required for setup or management, the lack of a common architecture, the lack of standardization among industries, and the lack of standardization with the farmer in mind as a user of ICT. Recently, several organizations have been working to develop standards, and some improvements have already been developed or are in process.

Centers that store machine, agronomic, and social knowledge will aggregate data to provide value-added services for machinery operation and farm management. Some of these data may be collected by farmers, and some will be provided by public and private sources of agricultural information. Some data sources, such as remote sensing, have been mentioned, but a number of others will emerge as the aggregated knowledge in efficient production agriculture increases.

Centers with machine knowledge can help increase equipment uptime and anticipate machine system failures based on vehicle state variables in operation. Machine data that provide a better understanding of machine use can also lead to more efficient system designs that meet the needs of farmers. Agronomic data will create new opportunities for intensive modeling and simulation that can improve production efficiency by anticipating the impact of weather and various production methods.

In the future, ICT will enable the development of new platforms that can provide more support to production agriculture by taking advantage of opportunities to connect farmers, the value chain, and society in ways that are beyond present capabilities. The German-funded iGreen project, for example, is working on location-based services and knowledge-sharing networks for combining distributed, heterogeneous public and private information sources as steps toward future ICT systems (iGreen, 2011). Today, we are extremely close to having true CPS and control systems for measuring the "pulse" of agricultural productivity on planet Earth.

INVESTMENT IN MECHANIZATION

There has been a substantial global investment in agricultural mechanization and automation by governments, industry, farmers and international agencies. In general, the return on investment has been spectacular. In North America and Europe, the combination of advanced mechanization systems, agrochemical inputs and plant breeding has produced an increase in farm production of such proportions that ultimately quotas on production had to be imposed to prevent the accumulation of massive food surpluses.

The lesson from this experience is abundantly clear. Food production can be increased if the primary producer or farmer is provided with a guaranteed profitable income for the farm produce. With such guarantees the farmer can invest in the necessary inputs including mechanization to increase productivity, secure in the knowledge that, as productivity increases, income will increase enabling payback and facilitating further investment as required.

The scale of the investment required may be determined from comprehensive data on farm mechanization costs (see also Expenditures and Returns, EOLSS on-line). For instance, the current cost of mechanization in the UK is about 20 % of total farm input costs.

SELECTION AND OPERATION OF EQUIPMENT

The choice or selection of agricultural equipment is dictated by a multiplicity of factors including the nature and size of the enterprise, the profitability and access to finance, the economic status of the region, the accessibility to a range of equipment options at local level, the ownership (individual, shared or cooperative) of equipment and access to mechanization contractors. For individual farmers in the developed world, a tractor is likely to be the key item of equipment as it provides power and mobility for a wide range of mechanical farm operations including tillage, spraying, fertilizing, harvesting, milking and feeding. The size and number of tractors is dictated by the size, nature and profitability of the farming enterprise as well as by how many (if any) operations (e.g. plowing, planting, spraying, harvesting) will be serviced by local contractors.

For example, the capital cost of purchasing a harvester to harvest a small area of a moderate value crop is often prohibitive. On the other hand, a contractor servicing many small growers in a local region can spread the capital cost accordingly and provide a service at an affordable price. For individual farmers who cannot afford a new and expensive item of equipment and who prefer not to depend on contractor availability, purchase of second hand equipment is an option where the support of a local finance agency may be required. Leasing of expensive equipment is also an option but generally is less popular except with larger producers.

In the developing countries, the trend towards adoption of the tractor as the fundamental unit of agricultural mechanization systems is sure to continue. However, for many small farmers in deprived regions the transition to tractor-based mechanization is not a realistic option due to the lack of finance and basic infrastructure. In this situation, a continued reliance on human energy and animal-powered equipment for tillage, planting and harvesting will prevail for the foreseeable future. As a consequence, the design and operation of such equipment requires increased investment in research and development to enhance their operational and performance characteristics.

NEW TECHNOLOGY AND THE FUTURE

The basic technology of agricultural machines has changed little through the last century. Though modern harvesters and planters may do a better job than their predecessors, the combine of today (costing about US$250,000) cuts, threshes, and separates grain in essentially the same way earlier versions had done. However, technology is changing the way that humans operate the machines, as computer monitoring systems, GPS locators, and self-steer programs allow the most advanced tractors and implements to be more precise and less wasteful in the use of fuel, seed, or fertilizer. In the foreseeable future, some agricultural machines may be made capable of driving themselves, using GPS maps and electronic sensors.

Even more esoteric are the new areas of nanotechnology and genetic engineering, where submicroscopic devices and biological processes, respectively, may be used to perform agricultural tasks in unusual new ways.

Agriculture may be one of the oldest professions, but with the development and use of agricultural machinery, there has been a dramatic drop in the number of people who can be described as "farmers."

Instead of every person having to work to provide food for themselves, less than two percent of the United States population today works in agriculture, yet that two percent provides considerably more food than the other 98 percent can eat.

It is estimated that at the turn of the twentieth century, one farmer in the United States could feed 25 people, whereas today, that ratio is 1:130. (In a modern grain farm, a single farmer can produce cereal to feed over a thousand people.) With continuing advances in agricultural machinery, the role of the farmer will become increasingly specialized.

Performance of Agricultural Equipment

Maintaining working conditions and optimal performance of agricultural equipment is of vital importance in agricultural and food production due to the timeliness factor. The concept of timeliness recognizes that there is an optimum time to perform certain crop production operations from planting through to harvesting. If one or more of these operations is performed too early or too late, a timeliness penalty is likely to accrue, that is, yield and/or the quality of the crop is diminished, yielding a lower price to the farmer.

For example, the ideal time to harvest grain is when the crop is ripe and the moisture is low (see also Maintaining Working Conditions and Operation of Machinery, EOLSS on-line). However, if the weather is bad and the quality of the grain is in danger, it may be more economical to protect quality by harvesting early (before ripeness) even if an increased post-harvest drying cost is incurred.

Due to the timeliness factor, machinery of a somewhat higher capacity is often employed to avoid timeliness penalties that may accrue due to the use of a contractor, machine breakdown and repair, bad weather or operator illness. Likewise, the use of preventive maintenance protocols are desirable in the off-season e.g. replacing parts (such as bearings or soil engaging elements) that may break or wear out before they need repair or replacement.

The use of highly skilled and competent operators is also desirable to ensure optimal performance of equipment that is generally getting bigger, working faster and becoming more complex despite the welcome introduction of more automated control and ergonomic systems designed to assist and enhance performance. The training of operatives involves a partnership between equipment users, equipment suppliers, maintenance and repair services, extension services (where available), research and educational institutions, and the communications media (farming press, radio, TV, Internet).

Farm relief services are an integral part of a back-up system where illness or other difficulties prevent a farmer from operating equipment effectively.

Agricultural engineers

Agricultural engineers work in the context of agricultural production and processing and the management of natural resources. Their specialties include power systems and machinery design; structures and environmental science; and food and bioprocess engineering. They perform tasks such as planning, supervising, and managing the building of dairy effluent schemes, irrigation, drainage, flood and water control systems. They develop ways to conserve soil and water and to improve the processing of agricultural products. In addition, they may perform environmental impact assessments and interpret research results.

FARM EQUIPMENT

Fig. 1: A modern Farm Tractor plowing a field using a chisel plow.

Fig. 2: A plow in action. Notice the soil being turned over.

Fig. 3: A round baler in action.

Traction and power

- Tractor
- Crawler tractor / Caterpillar tractor

Soil cultivation

- Cultivator
- Chisel plow
- Harrow
- Spike harrow
- Drag harrow
- Disk harrow
- Plow
- Power tiller
- Rotary tiller
- Rototiller
- Spading machine
- Subsoiler
- Walking tractor

Planting

- Broadcast seeder (or broadcast spreader or fertilizer spreader)
- Plastic mulch layer
- Potato planter
- Seed drill
- Air seeder
- Precision drill

- Transplanter
- Rice transplanter

Fertilizing and pest control

- Fertilizer spreader (see broadcast seeder)
- Terragator
- Manure spreader
- Sprayer

Irrigation

- Center pivot irrigation

Harvesting / post-harvest

- beet harvester
- Bean harvester
- Combine harvester
- Conveyor belt
- Corn harvester
- Cotton picker
- Forage harvester (or silage harvester)
- Huller
- Potato digger
- Potato harvester
- Sickle
- Swather

Hay making

- Bale mover
- Baler
- Conditioner
- Hay rake
- Hay tedder
- Mower

Loading

- Backhoe
- Front end loader
- Skid-steer loader

Other

- Grain auger
- Feed grinder
- Grain cart
- Rock picker

14

GLOBAL POSITIONING SYSTEM (GPS) AND GEOGRAPHIC INFORMATION SYSTEMS (GIS) FOR AGRICULTURAL AUTOMATION

These technologies enable the coupling of real-time data collection with accurate position information, leading to the efficient manipulation and analysis of large amounts of geospatial data. GPS-based applications in precision farming are being used for farm planning, field mapping, soil sampling, tractor guidance, crop scouting, variable rate applications, and yield mapping. GPS allows farmers to work during low visibility field conditions such as rain, dust, fog, and darkness.

In the past, it was difficult for farmers to correlate production techniques and crop yields with land variability. This limited their ability to develop the most effective soil/plant treatment strategies that could have enhanced their production. Today, more precise application of pesticides, herbicides, and fertilizers, and better control of the dispersion of those chemicals are possible through precision agriculture, thus reducing expenses, producing a higher yield, and creating a more environmentally friendly farm.

Precision agriculture is now changing the way farmers and agribusinesses view the land from which they reap their profits. Precision agriculture is about collecting timely geospatial information on soil-plant-animal requirements and prescribing and applying site-specific treatments to increase agricultural production and protect the environment. Where farmers may have once treated their fields uniformly, they are now seeing benefits from micromanaging their fields. Precision agriculture is gaining in popularity largely due to the introduction of high technology tools into the agricultural community that are more accurate, cost effective, and user friendly. Many of the new innovations rely on the integration of on-board computers, data collection sensors, and GPS time and position reference systems.

Many believe that the benefits of precision agriculture can only be realized on large farms with huge capital investments and experience with information technologies. Such is not the case. There are inexpensive and easy-to-use methods and techniques that can be developed for use by all farmers. Through the use of GPS, GIS, and remote sensing, information needed for improving land and water use can be collected. Farmers can achieve additional benefits by combining better utilization of fertilizers and other soil amendments, determining the economic threshold for treating pest and weed infestations, and protecting the natural resources for future use.

GPS equipment manufacturers have developed several tools to help farmers and agribusinesses become more productive and efficient in their precision farming activities. Today, many farmers use GPS-derived products to enhance operations in their farming businesses. Location information is collected by GPS receivers for mapping field boundaries, roads, irrigation systems, and problem areas in crops such as weeds or disease. The accuracy of GPS allows farmers to create farm maps with precise acreage for field areas, road locations and distances between points of interest. GPS allows farmers to accurately navigate to specific locations in the field, year after year, to collect soil samples or monitor crop conditions.

Crop advisors use rugged data collection devices with GPS for accurate positioning to map pest, insect, and weed infestations in the field. Pest problem areas in crops can be pinpointed and mapped for future management decisions and input recommendations.

The same field data can also be used by aircraft sprayers, enabling accurate swathing of fields without use of human "flaggers" to guide them. Crop dusters equipped with GPS are able to fly accurate swaths over the field, applying chemicals only where needed, minimizing chemical drift, reducing the amount of chemicals needed, thereby benefiting the environment. GPS also allows pilots to provide farmers with accurate maps.

Farmers and agriculture service providers can expect even further improvements as GPS continues to modernize. In addition to the current civilian service provided by GPS, the United States is committed to implementing a second and a third civil signal on GPS satellites. The first satellite with the second civilian signal was launched in 2005. The new signals will enhance both the quality and efficiency of agricultural operations in the future.

GPS TECHNOLOGY

Global Positioning System (GPS) receivers provide a method for determining location anywhere on the earth. Accurate, automated position tracking with GPS receivers allows farmers and agricultural service providers to automatically record data and apply variable rates of inputs to smaller areas within larger fields.

A GPS receiver can be compared with a simple AM or FM radio. A GPS receiver "listens" for the signals that are broadcast from the satellites of the United States Department of Defense (DOD) Global Positioning System. Orbiting around the earth at an altitude of 12,550 miles, these satellites are in predictable locations; hence, we refer to the system of satellites as the GPS constellation.

Each satellite broadcasts almanac information containing the position of all satellites in the constellation. GPS receivers use the almanac to determine the position of the satellites. Minor variations in the orbits of the satellites occur due to gravitational forces from the sun and the moon. The DOD continuously monitors the satellites and adjusts the almanac information to represent the actual orbits of the satellites.

The broadcast signals also contain a precisely timed predictable code that a GPS receiver can use to determine how long the signal required to reach the receiver. A microprocessor within a GPS receiver uses these delays and the position of the satellite to calculate the distance to each satellite, and then uses this information to determine location through triangulation.

Triangulation is a mathematical method for locating points on a plane in three-dimensional space. If the distances to each of three satellites and your approximate location on the earth are known, the GPS receiver can calculate its terrestrial position. If information from four satellites is available, elevation can also be determined.

Range determination factors

Each GPS satellite continuously broadcasts two radio signals on separate L-band frequencies (the L-band is from 1,000 to 2,000 MHz). The L1 signal (transmitted at 1575.42 MHz) carries two codes, a Coarse/Acquisition (C/A) code and a Precision (P) code.

The L2 signal (transmitted at 1227.60 MHz) carries only the P code, which is encrypted so only the military and other "authorized" receivers can interpret it. The use of both the L1 and L2 signals and their P codes produces what is called the Precise Positioning Service (PPS) and is available to the U.S. and allied military, U.S. government agencies and authorized civilian users. The system available for all civilian use accesses only the L1 signal and the C/A code and is known as the Standard Positioning Service.

Accuracy

The accuracy obtained generally depends on five factors.

- Proper installation
- The degree of technology used in the receiver
- The number and location of satellites
- Errors introduced by selective availability (SA), atmospheric conditions, the troposphere, the ionosphere, and multipathing — radio signals bouncing off objects in the area
- Differential corrections.

Installation

GPS antennas should be mounted on the centerline of a combine, tractor or truck and above any part of the machinery that might obstruct a line of sight to a satellite. If the cab is centered and the top of the cab is above other portions of the machine, a cab-top mounting may be the best location. However, on a steep side slope, a high mounting point will result in an error in position calculation due to the offset in horizontal position.

Example

If the antenna on a sprayer traveling 10 mph is mounted 30 feet ahead of the booms and a rate change at the controller is effected at the boom two seconds later, the rate change will occur when the booms reach the location of the antenna where the change was made.

At any other ground speed, the rate change at the booms will not occur at the same location as the controller. A time adjustment must usually be factored into the system to compensate for time delays in sensing or product application.

Electrical interference can result from electrical storms, power lines, 2-way radios, nearby radio transmitters, electric motors, microwave towers, cellular phones, vehicular electrical equipment such as alternators and ignition systems on spark-ignition engines, and other sources.

Changing the position of the antenna or adding noise suppression kits may reduce interference problems from alternators and ignition systems. Follow the instructions for installation of the GPS equipment, making sure that all connections are tight.

Technology

Low-cost receivers receive signals from one satellite at a time and require more time to determine the location than a receiver capable of receiving four signals simultaneously. Usually, seven to 10 satellites are in view at any one time and more sophisticated receivers will produce the most accurate location.

Reacquisition time is the time it takes to get an accurate position fix after a short-time loss of satellite signals; this may occur for a variety of reasons, including traveling near trees or buildings and losing the 'line of sight' to satellites. Reacquisition time is important for most agricultural applications and, especially, for guidance with applicators and aircraft.

New technology in GPS receivers has shortened reacquisition time. Receivers that can track 8 to 12 satellites are less susceptible to acquisition loss.

Satellite constellations

Using triangulation to calculate position, small errors in distance can cause large errors in position. The error in calculating position through triangulation increases when the satellites are close together. The best accuracy is produced when the receiver can pick up signals from many widely dispersed satellites (Fig. 1).

Selective availability and other errors

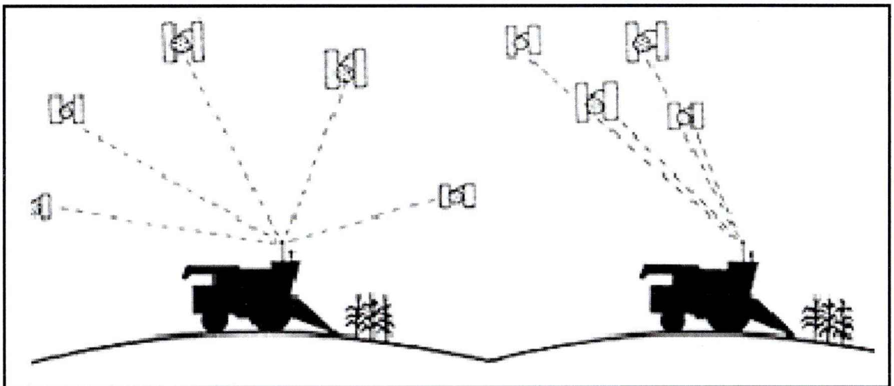

Good satellite geometry	Poor satellite geometry

Fig. 1: Good satellite geometry (widely and evenly spaced satellites) yields more accurate GPS position estimates.

To prevent an enemy from using GPS satellite signalsfor determining locations on earth, the DOD "scrambles" the signals sufficiently to introduce an error of about 100 meters in an uncorrected location calculation.

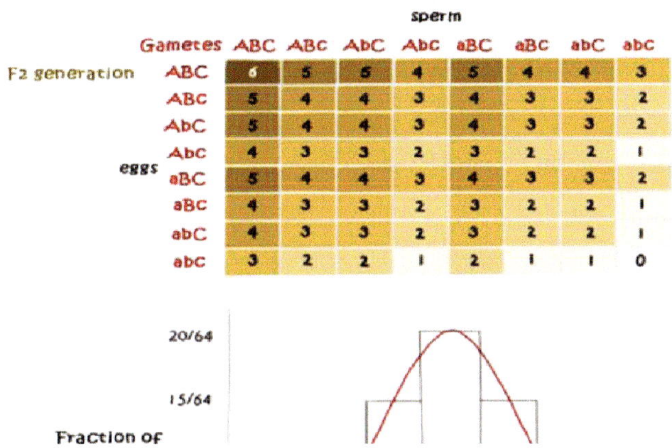

Fig. 2: GPS signals can somtimes "bounce" off of objects in their path and cause the signal to reach the receiver on a different path. This is called multipath error.

The term for this is "selective availability" (SA). Atmospheric, tropospheric and ionospheric conditions, however, also cause distortions or errors in calculating distance; natural errors due to these conditions are not easily or reliably predicted. Hence, even in the absence of SA, differential corrections will still be required to accurately calculate position.

Multipathing, the phenomenon that creates distorted television signals is caused by signals that bounce off of other objects before reaching the antenna (Fig. 2). Multipathing cannot be corrected by differential corrections.

Differential corrections

Stationary GPS receivers are used to calculate the total error due to SA, variable atmospheric conditions and other factors. The concept is simple. A stationary receiver always has a known location; because the actual positions of the satellite and the receiver are known, the true range (distance) is known. The distance calculated by the receiver using the broadcast signals is known as the pseudorange, which is generally in error due to the combined sources of all errors. The difference between the true range and the pseudorange is the error and is known as the differential correction.

Differential correction data can be purchased and used at a later time in a process known as post processing to correct the errors in recorded data. However, the most common approach is to connect a differential corrections receiver to a GPS receiver to provide real-time corrections (Fig. 3).

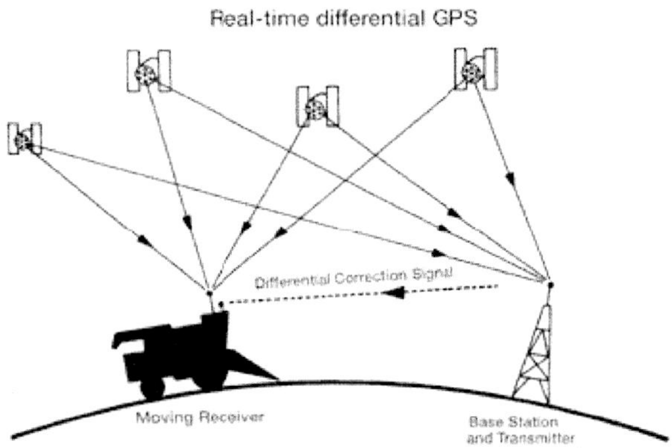

Fig. 3: In real-time DGPS, the stationary receiver transmits the differential correction to the moving receiver via another radio signal.

Many units incorporate GPS receivers and differential corrections receivers into the same unit. These are often referred to as differentially-corrected GPS (DGPS) receivers. Differential corrections signals are available from the Coast Guard or Army Corps of Engineers and through commercial sources, which, for a fee, will provide signals from a satellite or a land-based tower. Where these sources aren't available, or for special applications, a private differential corrections source can be installed.

Some of the newer DGPS receivers combine the capability of receiving differential signals from both the Coast Guard beacons and from a satellite service. Refer to Table 1 for a comparison of features of Coast Guard and satellite-based differential corrections sources.

Coast Guard signals

The Coast Guard signals are broadcast in the frequency range of 285 to 325 kHz (just below the usual AM-radio band) where radio waves travel as ground waves and are not limited to line-of-sight reception like FM-radio stations.

The signals are series of pulses similar to those of the GPS satellites. Referred to as Minimum Shift Keying modulation, the signal is less sensitive to electrical interference and noise than AM-radios.

Missouri has free access to correction signals from Coast Guard beacons located near St. Louis (@ 322 KHz), Kansas City (@ 305 KHz), Tulsa (@ 299 KHz), Rock Island (@ 311 KHz), Memphis (@ 310 KHz) and Omaha (@298 KHz).

The range of the Coast Guard beacons is approximately 150 miles in good weather (electrical storms cause interference). Accuracy decreases with distance from the transmitter. This service is expected to become the choice of many agricultural users, especially in Missouri where several signals are available.

A disadvantage of the Coast Guard differential corrections signal is the rate at which the beacon transmits or repeats messages. Most Coast Guard sites broadcast at 200 bits per second. At this broadcast rate, the age of a satellite's differential correction can be as old as four seconds. For some applications, such as guidance, this update rate may be unacceptable. For guidance applications, update rates of two to ten times per second may be required.

Typical Coast Guard beacon receivers have two channels. One channel receives the differential correction and the other is searching for the best incoming signal. This helps to ensure against loss of a DGPS signal if at least two beacons are within range.

Satellite-based correction signals

For the user, one of the simplest types of differential corrections signals is transmitted from a geostationary satellite. Companies such as Omnistar, Accqpoint and Racal provide this service.

The typical annual user's fee ranges from $500 to $800. The correction signal is available throughout most of North America. The accuracy of high quality receivers is generally considered to range from one to three meters RMS. Interference from man-made sources is minimal.

Satellite-based signals may have an advantage for operation around trees and buildings since the satellite is nearly overhead at most locations and within the DGPS receiver's line-of-sight.

Land-based correction signals

Several commercial land-based correction signal services are also available for a fee. Some companies put up their own transmitters to broadcast correction signals; these include SatLoc, Mobile Data and CSI.

Some commercial service providers piggyback correction signals onto commercial FM radio station transmitters. These sub-carriers include Pinpoint Communications, DCI and others.

Private GPS receiver and radio transmitter

GPS users not covered by Coast Guard or commercial sources of differential corrections can install a stationary receiver and transmitter to provide their own differential corrections source. Few users in Missouri will choose to buy and install their own fixed GPS receiver and transmitter since the Midwest has other choices available.

Cost vs. accuracy

The accuracy attainable with GPS depends partly on how much you are willing to spend, ranging from approximately $100 to $100,000. A low-cost (from $100 to $500) GPS receiver without DGPS capability may be sufficiently accurate for some crop scouting applications, for navigating highways or for locating your favorite fishing spot on a lake. The RMS horizontal accuracy may be about 50 yards.

The cost for a basic DGPS receiver suitable for most agricultural applications is about $3,000 to $5,000 and provides RMS accuracy of at least three meters with a typical accuracy of one meter, which is sufficient for yield monitoring and grid soil sampling.

If you need a GPS receiver for guidance (for spraying, fertilizer application, etc.), the cost may be up to $25,000. Such systems provide accuracy down to a few inches. Since sprayers and fertilizer spreaders can travel fairly quickly, lower quality GPS equipment may not update position quickly enough to be used for guidance or control, although GPS systems with high update rates and accuracies in the range of one foot or less are becoming available at lower prices.

The annual subscription cost for some differential correction services varies with the level of service (accuracy). Some providers offer three levels of service, e.g., one provider has a premium service for better than 1 meter accuracy, intermediate service for accuracies in the range of 5 meters and a basic service for accuracies in the range of 10 meters. Typical approximate costs may be $600, $250 and $75 per year, respectively, depending on the level of service.

Accuracy of GPS units

Accuracy of GPS units may be stated in a variety of statistical terms, which one is used may not be specified. Most statistical definitions of GPS accuracy assume that position errors are random in nature and follow a normal distribution. One measure of accuracy is the Circular Error Probable (CEP). This term applies to horizontal position estimates. A CEP of 1 meter is interpreted to mean that 50 percent of the position estimates will be within 1 meter of the actual position; the other 50 percent can be anywhere in the universe.

Two frequently used accuracy terms are the RMS (also known as one sigma) and 2DRMS (also known as two sigma). The acronym RMS stands for root mean square and is approximately equal to the standard deviation (SD). If the calculated positions were normally distributed about the true position, then 68 percent of the computed positions would be within +/- one standard deviation and 95 percent would be within +/- two SD's of the true position (2DRMS).

When comparing the accuracy of GPS units, make sure that the accuracies are specified in the same terms (CEP, RMS or 2 DRMS).

Coordinate systems

Several coordinate systems are in use for mapping and may cause problems with compatibility between software systems. Users frequently need to transform position data into a plane (flat) coordinate system, either to merge them with another data set, to plot a map of the GPS results, or to perform further calculations for such parameters as area, distance or direction (plane coordinate systems are usually easier to work with than geodetic coordinates). When using data and maps from several sources, coordinates must be based on the same datum. The coordinate system differences, which are caused by a different reference frame, ellipsoid and data adjustment, are significant (up to several hundred meters) and cannot be ignored.

Several commercially-available software programs produced by well-known GIS vendors treat the coordinate shifts incorrectly. The National Geodetic Survey provides software (LEFTI and NADCON) at a nominal charge to compute datum shifts. Boundary coordinates on older paper copies of soil maps should be converted to the preferred datum (probably WGS84) before they are digitized.

GPS receivers can usually report position information in more than one format. The most common format is lat/lon (latitude and longitude). Lat/lon coordinates are recorded in angular units of degrees, minutes and seconds. One second of latitude is equal to about 30 meters. GPS receivers may display lat/lon in degrees plus minutes to four decimal places (instead of minutes and seconds). Most geographic information system (GIS) software is capable of using more than one format and may automatically convert lat/lon coordinates to a coordinate system such as Universal Transverse Mercator (UTM) or State Plane Coordinates (SPC) to calculate distances in meters or feet.

UTM and SPC systems project portions of the earth's curved surface onto a flat map and report locations as actual distances from a reference point in meters and feet, respectively. Hence, no conversions are necessary to calculate distance or area.

Commercial software available from several GPS vendors will compute UTM or state plane coordinates from GPS data. These coordinates are usually based on the WGS-84 datum and thus are in the NAD-83 system. If these must be transformed to NAD-27, it is advisable to do the NAD-83 to NAD-27 transformation in geodetic coordinates, and then make the conversion to plane coordinates as the final step.

Universal Transverse Mercator Coordinates

The UTM coordinate system is a worldwide system originally adopted by the U.S. military in 1947, and since has been widely used by civilian mapping in many countries. The UTM system is consistent throughout the world and one set of equations will allow calculation of coordinates at any location. The world is divided into 60 zones each spanning 6 degrees in longitude and extending north and south from a latitude of south 84

State plane coordinate system

In the 1930s, the U.S. Coast and Geodetic Survey established a plane coordinate system for each of the 48 states. One to five zones were established in each state with a Lambert Conformal or a Traverse Mercator projection. The specific projection and the size of the zone was selected to fit the geometry of the state and to keep distortions at or below one part in 10,000.

Farmers and agriculture service providers can expect even further improvements as the GPS-server.net tracking system continues to modernize.

Benefits

- Precision soil sampling, data collection, and data analysis enable localized variation of chemical applications and planting density to suit specific areas of the field.
- Accurate field navigation minimizes redundant applications and skipped areas, and enables maximum ground coverage in the shortest possible time.
- Ability to work through low visibility field conditions such as rain, dust, fog and darkness increases productivity.
- Elimination of the need for human "flaggers" increases spray efficiency and minimizes over-spray.

GIS IN AGRICULTURE

By mapping geographic and geologic features of current (and potential) farmland scientists and farmers can work together to create more effective and efficient farming techniques. Doing this could increase food production in parts of the world that are struggling to produce enough for the people around them.

GIS can analyze soil data combined with historical farming practices to determine what are the best crops to plant, where they should go, and how to maintain soil nutrition levels to best benefit the plants.

Many organizations are now implementing GIS systems including the USDA. They use many GIS variations in each of the USDA sectors to best capture what that department specializes in; in recent months, however, the benefit of combining this information has been realized due to the incredible capacity of GIS to transform and combine large amounts of data into a data set.

In the United States GIS systems are used by the USDA to protect crops, solve crop issues, and investigate fraudulent claims of crop damage as well as give farmers an easy way to access information about their crops season by season.

Agricultural Geographic Information Systems (AGIS) can map not only topography and crop health, but help solve wider economic issues in municipalities and urban centers that may stem from rural farming practices.

Precision Farming

Sensors on satellites, tractors and in fields are constantly collecting data. GIS and other technologies shape this data into information that is accessible and interpretable by farmers and land managers to make efficient and informed decisions.

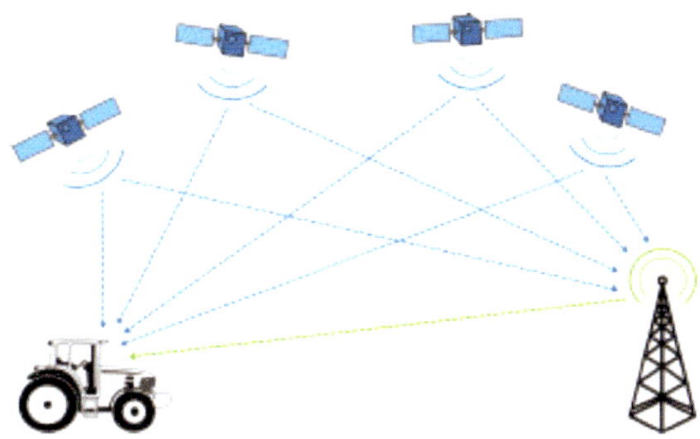

Fig. 4: Sensors on satellites, tractors

The use of precision GPS by farmers has become an essential tool in the agriculture industry. The technology has many cost saving and productivity advantages for farmers, as well as benefits for the community and the environment. A key example being its use in fertilizing crops, where machine sensors gather information about the crops, and the GPS records the exact position it is applied on the field. The technology then aids the application of fertiliser only to the areas where it is needed, and can vary the rate of application to target nutrient-deficient sites . This saves money on fertiliser product, as well as the environment from over application and runoff into local streams and rivers.

Real-Time Mapping

The use of satellites and drones has dramatically grown in the past few decades. With one of the key farming challenges being the availability and management of water for agricultural purposes, satellite technology can collect real-time data from the Earth's surface to assess and monitor the condition of the land. Soil Moisture Ocean Salinity (SMOS), vegetation growth using Landsat imagery, and applying the Normalised Difference Vegetation Index (NDVI), as well as a variety of other factors, can all be used to help estimate crop productivity and monitor drought and flooding on a global scale.

Drone technology is useful in collecting more local field data such as; plant height and biomass, flora counts, disease and weed presence, nutrient values, elevation and volume calculations. Drones can replace time-consuming tasks, normally performed by scouring fields on foot, to assess qualities of the flora and vegetation present. Farmers are thus able to cover more ground, inspecting the health of the crops using aerial imagery and other data that the drone can record.

For these purposes, and many more, satellites and drones can be used by farmers to make real-time actionable decisions. Monitoring yields, applying nitrogen, using precision water sensors and identifying critical areas for intervention are all valuable uses that this technology can provide. Just one basic drone can provide farmers with the ability to make many powerful decisions that will impact the health and productivity of their land.

In the United States of America, the National Agricultural Statistics Services (NASS) have developed an online mapping application called CropScape. This application can provide area estimates of crops, estimate the type of crop growing and how large the size of its yield could be. Not just for privatised use, government agencies have used CropScape data to assess such national issues as; food security, pesticide control and changes to land-use.

Meeting the Future Demand for Food

For today's governing agencies, the need to secure the necessary food sources to support the continuous rise in population has been a growing concern. It has been estimated that the current production of crops will need to double by 2050 in order to meet future needs for food. For such purposes, GIS is not only being used for real-time analyses, but also to compare historical data. Landsat satellite imagery can be used to assess historical trends of agricultural land use over time. This can help to predict and plan for the amount of arable land needed to supply food to the future populations.

Raising Awareness

One of the most important functions of GIS and mapping is to be used to raise awareness of such issues as food scarcity, and locating areas that are in need of assistance. Web mapping tools like the Institute of the Environment and the University of Minnesota's Feeding the World map, and the Food and Agriculture organisation of the United Nations (FAO)'s World Hunger Map provide a unique view of global food production. Though establishing the underlying causes of food insecurity, GIS data and technology are helping to safeguard the areas and communities affected by food scarcity.

The agriculture industry has been greatly transformed by geospatial technology. In just the past few years, the online (and offline) power and accessibility of GIS and digital mapping both has changed the way farmers and managing their land and performing their most basic tasks (Takor Group, 2017).

GIS allows us to visualise, analyse and understand that vast extent geographic data that is being stored and collected on a daily basis. It can now be easily communicated to us which crops are flourishing, the extent that pollution and natural disasters hinder production and control the use of fertilisers.

Crop yield prediction

Accurate yield prediction can help governments ensure food security and businesses forecast profits and plan budgets. The recent development of technology connecting satellites, sensing, big data, and AI can enable those predictions.

One of the most profound techniques in this field is Convolutional Neural Networks (ConvNets or CNNs). A ConvNet is a deep learning algorithm that is taught to identify the productivity of a crop. Developers train this algorithm by feeding it images of crops whose yield is already known to find productivity patterns. CNN has an accuracy of about 82%.

A crop-prediction technique workflow

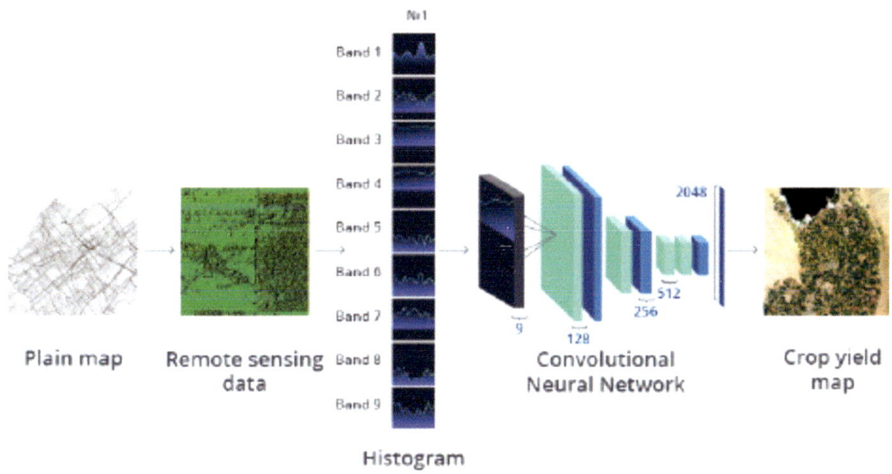

Fig. 5 crop-prediction technique workflow

Crop health monitoring

Checking crop health across multiple acres manually is the least efficient option. This is where remote sensing combined with GIS in farming comes to the rescue.

Satellite images and input information can be paired to assess environmental conditions across the field, such as humidity, air temperature, surface conditions, and others. Based on GIS, precision farming can upgrade such an assessmen and help you decide which crops require more attention.

A more sophisticated approach uses imaging sensors on satellites and air vehicles to check the temperature of crops. When the temperature is above normal, this might indicate a disease, infestation or insufficient irrigation.

Neural networks like CNN, Radial Basis Function Network (RBFN), Perceptron, and others can be helpful in assessing crop health too. The algorithms can analyze images for unhealthy patterns.

Livestock monitoring

The simplest application of farm GIS software in animal husbandry is the tracking of movement of specific animals. This helps farmers find them on a farm and monitor their health, fertility, and nutrition. GIS services that allow you to do that comprise trackers installed on animals and a mobile device that receives and visualizes information from those trackers.

Here's one example. You want to monitor the weight of your beef cattle. Each animal has a tracker on its ear or neck. Every time it steps on the digital scales, the scales read the ID of that animal and assign a new value to that ID in the system.

You don't need to manually enter that data. Meanwhile, if there's an alarming change in the animal's weight, you can quickly find that animal and check its health.

There are also more interesting use cases of farm GIS software, such as preventing wolf-cattle encounters. There are ambiguous spatial specifics that affect the distribution of wildlife in an area, including wolves. We could reduce undesirable encounters by understanding those subtle specifics, which could be done by the combined use of AI and GIS in agriculture.

Insect and pest control

The invasion of harmful insects and pests, or infestation, does heavy damage to agriculture. A look from above can enable accurate, timely alarms to prevent that.

Yet even high-resolution images might not provide visible early signs of infestation.

The alternative would be using AI. You develop a neural network and train it using deep learning algorithms. Through this training, you feed the neural network images of infested land, and the network learns to find samples that indicate infestation. After that, you feed it satellite images of the land you want analyzed.

As mentioned above, you can also use remote sensing along with geospatial technology in agriculture to check the temperature of the crops. Plants respond to infestation by heating up as they stop getting enough water or nutrition.

Irrigation control

Keeping an eye on vast fields to make sure that each crop gets enough water is a challenging task, but one easily tackled by geoinformatics in agriculture.

Aircraft and satellites equipped with high-resolution cameras take images that allow AI algorithms to calculate the water stress in each crop and spot visual patterns behind water shortages.

Pair those images with water delivery system maps, and you will find out how well your current irrigation scheme is performing.

Flooding, erosion, and drought control

Marrying GIS and agriculture can help prevent, assess, and mitigate the negative impact of destructive natural phenomena.

To identify flood-susceptible areas, you can use flood inventory mapping techniques. You need to collect data such as past floods, field surveys, and satellite images. Use those data to create a dataset to train a neural network to spot and map flood risks, and you will create an ultimate disaster management tool.

If you need to check land for susceptibility for soil erosion, you could pair Universal Soil Loss Equation (USLE) with GIS and remote sensing. Run satellite images through spectral analysis to check USLE factors and verify those images with field observations. As a result, you can create a map featuring the level of deterioration of the soil across the field.

Similar GIS solutions for agriculture can be used to control drought.

Farming automation

Seeding machines, intelligent irrigation systems, driverless harvesters, and weed remover robots are the inevitable future. You could equip each of your machines with sophisticated sensors, but why do that if you can connect them to an integral GIS system?

GIS in farming can provide precise maps, including all necessary information about the crops in the field. Maps like those are called task maps or application maps. Smart machines use them to tend to the field.

Here's an example of how GIS solutions for agriculture might work. Once a GIS system has detected weed infestation, it assigns a "Weeding needed" label to that area. The weed remover robot reads the label and places this area on its list of tasks.

Apart from providing signals for machines, task maps can help unskilled workers do their job more efficiently.

Benefits of GIS in agriculture

Remote Sensing and GIS technology enable agencies to get reliable information of natural and man-made features or processed and interpreted appropriately phenomenon occurring over the earth's surface without making any physical contact. Nowadays, farmers rely on these technological improvements.

- **Crop Sown Area Estimation**: Crop sown area estimation is one of the major sections in agriculture Remote Sensing activities. Remote Sensing plays a crucial role in mapping and monitoring various crops sown area estimation. Satellite data such as Sentinel-1, 2, Landsat-8, World View-3, LISS-IV, and more provide precise level crop sown area and helps in crop loss assessment due to various catastrophic disasters.

- **Normalized Difference Vegetation Index (NDVI)**: This is primarily used for the assessment of vegetation dynamics, particularly in determining the crop health status. With NDVI, the possibility of understanding the crop phenology escalates as it explains the crop chronology and their relationship with weather and climate (season). NDVI is measured using mathematical calculation of spectral bands within the satellite image which measures the healthiness of vegetation, as it has a robust correlation with green biomass indicating healthy vegetation or crop.
- **Crop Diseases Identification:** Remote Sensing technology provides spatially distribution of information of diseases and pests over a large area with relatively low cost. With the help of satellite imagery and spatial analysis techniques, crop-infected areas are identified like Mealybug, Plant Hopper, and White Fly to get an overview for evaluating the potentially infected areas.
- **Soil Properties**: Soil properties are crucial in farm management practices as it has a direct impact on the yield output. Change in farming system and land management results in soil change which compromises the current and future capacity for primary production and provision of crops and by micronutrition mapping as the soil is a very important aspect of agriculture with characteristics like soil pH, soil organic matter, soil texture among many others. These characteristics infer information about soil condition by observing what happens on the surface in terms of vegetation growth. On the other hand, water content available within the soil is a very important factor that is taken into consideration by soil moisture mapping.
- **Flood Impact**: Every year in the Kharif season in India, the majority of the agricultural area is damaged due to flash floods or excessive rainfall. Satellite Remote Sensing provides significant information through the use of satellite imagery along with ground-based data collected from ground surveying teams, to compute precise damage assessment. In case of flood due to rainfall, the excess amount of water caused by the precipitation affects the agricultural area where the water drainage system is absent causing inundation. Flood due to overflow ultimately forms small tributaries and joins the river, creating a situation of flood in the vicinity of river banks and river plains. The damage assessment of floods can significantly improve the role of land use planning in managing flood risk.
- **NATCAT Modelling**: Natural Catastrophe modeling is a system to estimate the real-time or possible forecast of risk assessment, using the probabilistic approach to predict the outcome and behavior of natural hazards. This includes risk mapping and measuring hazards through computer-simulated catastrophe models where scientific studies and historical occurrences are linked with advanced information technology and geographic information system (using Remote Sensing & GIS). Flood risk maps are prepared depending on the estimated depth of inundation

of flood. The estimation is commonly derived from various hydrological and remotely sensed data where the process ensures that areas having a higher depth of inundation will be assigned a high 'hazard denomination'.

- **Drone Image Analysis for Crop Damage Assessment**: Drone image analysis is very useful in micro-level-crop assessment for crop loss due to hailstorm, horticulture tree counting, diseases, and many more. The accuracy of drone image data is directly related to the spatial resolution of the input imagery as the spatial resolution is very high ranging from 50cm and increase as per the requirement.

15

AGRICULTURE 4.0: STRATEGIC OPPORTUNITY AND CHALLENGES

Agriculture 4.0 is a term for the next big trends facing the industry, including a greater focus on precision agriculture, the internet of things (IoT) and the use of big data to drive greater business efficiencies in the face of rising populations and climate change. The digital transformation generated by information and communication technologies (ICT) affects every aspect of human society, enhancing traditional innovation and development processes. Moreover, it generates rapid product and process innovations that affect different segments of society (from the economy, communication, e-government, services, art, and archaeology to medicine and science).

These processes are based on the use of a series of enabling technologies, divided into product-service and process innovations, which are of strategic economic importance, including, in particular, the Internet of Things and big data. The Internet of Things encompasses the applications of digital technologies in everyday life (recording of habits and actions with sensors and/or devices; exchange, storage, sharing and processing of information and data through connection to the Internet). On the other hand, big data, also known as 'megadata', refers to the great speed with which data are generated, stored, processed, and analyzed (thus becoming a real resource to support decision-making and process automation).

NEW TECHNOLOGIES

New technologies are already disrupting the established norms of farming, with previously unaffordable devices now accessible and regularly deployed on farms across the world. Scout drones provide an "eye in the sky," scouting for pests in the field or dry spots requiring extra attention. The latest advances in sensor technology means drones are now able to use additional wavelengths in the light spectrum to assess crops, spotting weeds and sick crops from the air.

Blockchain is also on the rise, creating a new mode of interaction in the supply chain: trust is established by consensus between parties and backed by technology, eliminating the need for an intermediary.

It doesn't end there; blockchain has applications that go way beyond basic transactions. Blockchain can reduce inefficiency and greatly improve food safety. Traceability is also improved, with regulators quickly able to check the source of foods and detect the scope of any contamination issues.

These and other technological breakthroughs are acting as a significant disruptor, driving change and greater efficiency for hundreds of industries– including agriculture.

TECHNOLOGICAL RE-SCRIPTING OF SOCIETY

We do not dispute that smart farming technologies can play an important role in sustainable agricultural production, although the need for a technical revolution would be reduced if there were societal changes elsewhere.[2] The potential for improved productivity is likely to provide social benefits (e.g., greater food/income security), and environmental benefits, as less land is put into production. Precision agriculture, in combination with more productive crop varieties/livestock and the use of decision support systems to foster evidence-based decision-making, can lead to the smarter use of inputs with greater rewards. Furthermore, robotic technology could provide benefits to farming communities in compensation for lost labor, which is becoming a serious problem in the developing world[3] as the population migrates to urban centers. Technology-based SI could enhance social sustainability by supporting the profitability of farming businesses, and by providing different high-tech jobs.

Despite the potential benefits of a new technology revolution, the dominant techno-centric narratives associated with smart farming should be treated with caution. Technology is a double-edged sword because it has the potential to cause harm, as well as provide benefit. In agriculture, we have witnessed several controversies over the use of chemicals, including DDT and the ongoing issue of neonicotinoids, as well as intense debates over genetic modification. We have seen, for example, how previous technology revolutions have caused mass rural unemployment. The potential side-effects of smart technology like AI are being seriously considered now in policy.

Studies have shown that technology shapes agricultural society at several scales; on-farm, and across farming and non-farming communities argued that the requirement to use decision support tools would change, or "re-script," the ways in which farmers interacted with their land. Other studies have looked at the impacts of robotic milking technologies on-farm. Researchers argued that robotic technologies could change what it is to be a "good" farmer. The introduction of technologies to aid livestock management will mean that farming becomes less "hands-on" and this may change the nature of stockmanship and the relationship between animals and farmers argue that such technologies could create new rural subjectivities, in other words what it is to be a farmer or advisor. Negative impacts on-farm also influence imaginaries at wider scales.

Researchers argued that a focus on precision farming legitimates a narrative of chemical-based agriculture, and researchers suggested that the emphasis on big data could further move decision-making power from the farmers into the hands of private companies who have control over such data. In terms of wider impacts on society, the use of emergent technologies may not chime with societal expectations of sustainable food production. Using GM and other emerging technologies as analogs, it is likely that similar controversies will occur during this technology revolution if the views of publics are not adequately addressed. It is true that some smart technologies, such as precision agriculture, have so far been embraced with little societal "backlash," yet it is argued that large-scale use of AI, robotics, and other emergent innovations have the clear potential to cause unintended, unforeseen, and unwanted societal consequences. We have been here before.

Past controversies have reinforced the notion that science and technology are socially and politically constituted, and that emergent technologies can be met with societal resistance. Since there is often an institutional void for new technologies, it can be difficult to know how to govern implementation. Yet, past experiences highlight the peril of ignoring risks, the danger of becoming "seduced" by innovation, and the fallacy of not seeking the views of publics in an effort to construct a shared vision of the future.

TECHNOLOGY DRIVEN AGRICULTURE REVOLUTION

Agriculture is one of the oldest professions known to mankind. It began over 10,000 years ago across several areas of the world. Early humans were foragers or hunters – hunting wild animals and gathering wild fruits and roots. As humans evolved, so did farming. Agriculture as we know it today, has evolved remarkably from the labor intensive stage of the 1920s to the mechanized farming stage of green revolution, and finally the coming of age - Agriculture 4.0 which marked the adoption of digital technologies for increasing agricultural production and farm incomes.

The Agriculture sector is the largest livelihood provider globally. In India, despite contributing almost 15% to the GDP and employing over a quarter million people, the sector faces significant challenges, which hinders it from achieving its potential. The problems faced by the sector include:

- Farmers not getting the returns they deserve due to the presence of intermediaries
- Crops getting wasted due to insects and bug infestation
- Droughts affecting crop yield in one region while losing crops to flooding in another
- Small and fragmented farmlands losing more and more area to urbanization

- Migration to urban centers resulting in labor shortage at farms

Other factors such as climate change, population explosion and food security has forced the industry to seek innovative approaches that enhance production with lesser impact to the environment.

This is where "Agritech" or technology-driven agriculture comes into play. Agritech has unleashed a digital wave that has revolutionized the face of agriculture to deliver sustainable and profitable food production. The new era of agriculture - Agriculture 4.0 - is witnessing the augmentation of this manpower-driven system with Information and Communication technology (ICT) to achieve higher food production using fewer resources. Agriculture 4.0 marks the digitalization of food and agriculture systems using AI, IoT, automation and other technologies to create a hyper-connected network of farms, machines and factories that result in optimization of both food production and consumption.

Adoption of technology to enhance quality of produce

Urban Farming	Adoption of aeroponic technology for indoor vertical farming
Genetic Farming	Adoption of Clustered Regularly Interspaced Short Palindromic Repeats (CRISPR) for genome editing to create enhanced breeds with higher yields and resilience in adverse conditions
3D printing	Creating food objects using printers with hydrocolloids to replace base ingredient of food with other renewables like algae or grass

Adoption of cross-industry horizontal solutions

IoT	Using IoT for effective surveillance of farms with the help of sensors (light, humidity, temperature, soil moisture, health of equipment), automating irrigation systems, etc.
Drones	Utilizing drones for soil health scans, monitoring crop health, spraying fertilizers, forecasting crop yield and tracking weather conditions
Precision Farming	Big Data Analytics and Robotics-enabled approach that monitors farms and collects information like weather, yield, NDVI, crop moisture, terrain type, soil map from different sources to manage and optimize crop yield
Data Analytics	Making data driven decisions by analyzing data collected from IoT sensors at farms and the equipment used to deliver optimized smart farming
Nanotechnology	Application of novel nano tools like nanofertilizers, nanopesticides, nanobiosensors and nano-enabled remediation strategies for contaminated soils, focused on improving yields and food quality
Telematics	Machine to machine communication between hardware and sensors to automate decisions in farming. Application - precision- sowing that links seed spreader to soil sensors and leads to effective sowing as per soil conditions

Artificial Intelligence and Machine Learning	Application of autonomous robots to handle manual tasks, leveraging computer vision and deep learning algorithms for crop and soil monitoring, applying predictive analytics to track and predict environmental impact on crop yield
Big Data	Data mining large data sets generated by adoption of ICT in farming to provide actionable farming solutions that ensure profitability, efficiency and sustainability. Big data-enabled predictive modelling increasingly being used in crop insurance.
Blockchain	Enhancing agricultural supply chains by creating a shared, decentralized blockchain ledger of agricultural information that serves as a transparent and trusted source of truth. Empowers farmers and creates a direct link between farmers and consumers with traceability options

In recent times, the government has undertaken diverse digital initiatives such as Agricultural Technology Management Agency (ATMA), National Mission for Sustainable Agriculture (NMSA), E-NAM, and Soil Health Management (SHM), etc. that focus on uplifting the agriculture sector. In addition to govt. initiatives, India is home to more than 450 Agritech start-ups, growing at a rate of 25% annually. The sector has received more than $248 million in funding which clearly indicates the exponential potential of this sector.

However, despite all of these efforts by the government, the startups and the technology companies to revolutionize the Indian landscape using new business models, the technology adoption has been slow. The primary deterrent to widespread adoption is the perceived cost of technology adoption and functional illiteracy of the marginal farmers. In addition, small fragmented landholdings, high opportunity cost to reach the farmers as well as distortionary subsidy practices by the government have decelerated the momentum of Agriculture 4.0 in India.

The key to addressing the problems in digital adoption by the agriculture sector lies in Public Private Partnership (PPP) that can create an enabling environment for technologies to reach the deepest rural pockets in India and empower farmers at the grass-root level. The PPP should aim at customizing challenges locally, encourage innovation and entrepreneurship, support institutions and policies that create an enabling ecosystem and promote inclusive growth of farmers by harnessing the potential of these technologies. The agriculture revolution in India can only be brought about by a successful public-private partnership, with the private sector bringing it forward and the public sector creating an enabling environment for it.

BROADENING RESPONSIBLE INNOVATION FOR THE FOURTH AGRICULTURAL REVOLUTION

The concept of responsible innovation should underpin the fourth agricultural revolution; simply, ensuring that innovations designed to improve productivity and/or eco-efficiency also provide social benefits, meet human needs and are socially responsible. Noting that the concept is not well-developed in agriculture, researchers sought to develop a framework for responsible innovation in the context of smart dairy farming, while suggesting its possible applicability to other agricultural contexts build on the framework mechanisms to guide technology development organized around four dimensions—anticipation, inclusion, reflexivity, and responsiveness.

Researchers also argue that the responsible innovation process should seek to include all affected actors, such as technology companies, farmers, and local communities. Since sustainable food production has impacts across society, inclusion should be considered holistically. Indeed, we argue that inclusion should underpin all aspects of responsible innovation frameworks since anticipation, reflexivity, and responsiveness, cannot be undertaken without the inclusion of diverse voices.

In addition, researchers also suggested that agri-tech innovation projects should adopt structures to guide reflexivity, in other words there should be many opportunities to assess whether mutually beneficial trajectories are being followed. If not, then it is argued that innovation projects should be able to respond to challenges and be adaptable. Agri-tech innovators should also be able to respond quickly to problems caused by a new technology.

While these signs of progress are to be encouraged, and the recent interventions of researchers have made good progress in directly translating principles from the existing responsible innovation literature to smart farming, we argue that this agenda needs to go further. There are signs that responsible innovation is being taken more seriously in policy and funding circles. But, the specificities of the smart farming revolution create challenges for responsible innovation frameworks that have largely been developed in other fields, and have not been tested in practice in an agricultural context. For the purposes of the current discussion we highlight three main areas where we feel further work is needed.

RESPONSIBLE INNOVATION SYSTEMS

First, is the need to develop a more comprehensive and systemic approach to responsible innovation for the fourth agricultural revolution. The proposals by researchers mainly focus on possible tools and actions to enhance responsible innovation at the level of "R&D projects" and discrete technological innovations in the areas of smart dairying and smart agriculture. This partly reflects the frame of established responsible innovation frameworks which have been, for good reason, drawn to focus attention on controversial emerging technologies where one might expect to see significant negative consequences and questions of responsibility. We suggest, however, that in addition the frame of responsible innovation needs to be broadened.

We should ensure that the concept of responsible innovation is not attached solely to "big" emergent smart technologies. This would help to deliberately foreground questions over possible *alternatives* and the *directionality* of innovation pathways in addressing problems of sustainable agriculture, which have been shown time and time again to be associated with deeply held public concerns about emerging science and technology. In addition, a more systemic perspective would enable responsible innovators to consider interrelations between multiple co-existing innovations in sustainable agriculture. This could promote the cultivation of distributed responsibilities across wider innovation ecologies, as opposed to thinking about particular smart technologies in isolation. The more systemic thinking we advocate here demands methods capable of mapping multiple innovations—and tracking their emergence and interactions—across fields of innovation in sustainable agriculture. It also facilitates (and in turn demands) more strategic levels of governance and coordination between public, private, and civil society actors involved in steering innovations in the agri-tech revolution toward more socially responsible and humane ends.

BROADENING NOTIONS OF 'INCLUSION'

Second, there is a need to broaden notions of "inclusion" in responsible innovation to attend to the diverse ways that societal actors are already engaging with smart farming innovations on their own terms. This is a particular blind spot in frameworks for responsible innovation and an important omission in the proposals for "inclusion" put forward so far in relation to smart farming. Agricultural research is still dominated by top-down, non-inclusive approaches, and rarely includes relevant stakeholders, such as farmers, at an early stage. The power to shape research trajectories, including innovation design, rests with research institutions, and project narratives are rarely opened up in consultation with stakeholders. Responsible innovation encourages us to think of what technologies are for, who they are serving, and who is driving the process. Inclusion has sometimes been regarded as problematic by innovators, who see public involvement as potentially increasing time to market and releasing sensitive information publicly. But, research has also shown that open innovation and responsible innovation can be complementary.

In order to include all relevant actors and address the "gap in comprehensive inclusion" in smart dairy farming, researchers propose a range of tools and procedures including citizen forums, "workshops," "surveys," internet forums, interest group representation on steering committees and "user-centered design." Researchers echoes these procedures for including "rights holders" and advocates deliberative processes to open up democratic debate over the social and ethical implications of smart agriculture technologies. Such moves are important and necessary to open up inclusive debate and that might socially shape these emerging technologies. However, as researchers have argued, notions of participation and inclusion in responsible innovation need to go further. These existing proposals for smart agriculture fit in with what they identify as a dominant imagination of inclusion in science and innovation—i.e., a drive to invite all relevant actors into inclusive fora and collectives to socially shape specific projects or innovations that are the target for responsible innovation. Researchers suggest that while this is important, there is a need to also open up to wider "ecologies of participation" and already existing forms of societal engagement around these very same projects and innovations, which are legitimate sources of public concern, values, and actions that would otherwise be excluded from processes of responsible innovation.

Such moves resist new forms of inclusion in responsible innovation succumbing to an "public acceptance" model of innovation, where the onus is on innovators to change public opinion to accept technologies, rather than making technological trajectories more responsive to the needs of society. This calls for a new range of tools and methods in responsible innovation of smart farming and beyond, which are capable of mapping across diverse forms of societal engagement with emerging technologies and across innovation systems. This can draw on a number of promising mapping methods that are emerging in the social sciences and humanities, including digital methods (Rogers, 2013), issue mapping, systematic mapping and forms of comparative case study and meta-analyses.

Such methods can produce new forms of social intelligence about diverse forms of societal and farmer engagement with smart farming, for example: more formal spaces like farmer networking events, farmer clusters, demonstration farms, demonstration test catchments, and consultations, through to a multitude of informal spaces where farmers and publics are interacting with sustainable agriculture including interactions with vets, advisors, seed merchants, and livestock markets, discussions on social media, more low-tech forms of agricultural practice, community-based agri-environment solutions, and so on.

This forms a key part of a more comprehensive responsible innovation framework for the fourth agricultural revolution capable of mapping existing spaces of participation, finding out what innovations and ideas are being considered there, what alternative public concerns are being mobilized, and so on. This is not only a question of inclusion, but can also enhance reflexivity, anticipation, and responsiveness in responsible innovation frameworks. The field of sustainable agriculture can learn from other domains where such mapping approaches have been developed, and new institutional architectures of observatories have recently been proposed, for the responsible development of low carbon energy transitions and gene editing.

RESPONSIBLE INNOVATION IN PRACTICE

Third, research needs to assess whether responsible innovation frameworks make a difference in practice. This is a point that researchers makes in the context of genetic modification. He argues that responsible innovation frameworks must "demonstrate their capacities to shape existing technological trajectories." We find little, if any, evidence that responsible innovation frameworks have been tested in practice in the context of smart farming, before then being refined on the basis of feedback. If we do not know whether and how they affect the innovation process—in other words forcing innovators to *anticipate* problems caused by their product, to *respond* to emergent problems, to *include* diverse publics in an attempt to accommodate concerns, and lastly to adopt a *reflexive* and flexible approach to development—then they serve little practical purpose.

We argue, therefore, that the development of a refined responsible innovation framework for smart farming in the context of sustainable agriculture could be undertaken concurrently with practical trials. For example, prototypes of the framework could be introduced into innovation teams to see whether and how design practices are influenced, before then being refined based on observation and feedback from the innovators themselves. This iterative approach, which could also involve an assessment of how innovators themselves come to "know" diverse publics, would make a valuable contribution both to the broad field of responsible innovation as well as to the application of its principles to the fourth agricultural revolution.

TECHNOLOGIES USED IN AGRICULTURE 4.0

The use of new technologies in agriculture can be applied in various ways. Thus, these new methods can be used to collect data of interest, which allow analysis to be carried out for greater optimization in the use of available resources. On the other hand, it is also possible to use new technologies to achieve greater automation and efficiency when performing tasks in crops.

Some of the most important technologies and techniques used in agriculture 4.0 are explained below:

Drones

They are used to fly over crop fields. They are typically equipped with various **sensors** that allow them to **collect dat**a. One of the most common ways to use them is with a camera. Likewise, images of fields can be obtained from the air, from which various data such as soil fertility or crop growth can be extracted.

Fig. 1: Drone

Environmental sensors

They allow obtaining information about temperature, soil or humidity. With these data it is possible to know the needs that the crops may have at all times and take the appropriate measures. An example would be knowing the right amount of fertilizer for each crop. In this way, a greater optimization of resources is achieved.

Fig. 2: Environmental sensor

Cameras

Used to monitor crops and obtain information. There are different types such as conventional cameras or **Hyperspectral Cameras**, with which the composition of crops can be analyzed and data that is not visible with the human eye can be obtained.

Robotics

As in industry, through the use of robotic arms or **autonomous vehicles** in agriculture 4.0, it is possible to automate the performance of various maintenance tasks in the fields or to check the status of the crops through the use of other sensors, and take the necessary measures. For example, one of the possible applications of this technology would be to automate the pruning of vines.

Fig. 3: Camera

The data extracted by means of the new technologies can be processed using several already known techniques. In this way, it is possible to obtain relevant information in an automated way, which, otherwise, could go unnoticed by the human eye. The data obtained can be used in agriculture 4.0 with multiple purposes (soil study, execution of tasks ...). Some of the techniques that are possible to use are:

Computer Vision

With the images obtained by cameras, it is possible to apply different operations and processes with which to extract information that may be useful for the development of the crop, such as its size or the number of shoots.

Machine Learning

With the data collected through other technologies and sensors, it is possible to make use of **Machine Learning**. Using this technology, you can analyze the data obtained to draw various conclusions, while learning new information from these data. By using these techniques it is possible to determine, for example, the optimal time for harvesting crops, or the necessary measures for a higher quality of the product.

Data analysis

With the information collected with the sensors used, analyzes can be carried out that allow us to know the needs of the crop or future problems that may arise, such as pests and diseases. From this it is possible to take measures that may affect production.

ADVANTAGES OF AGRICULTURE 4.0

As already mentioned, the integration of new technologies with agriculture 4.0 offers a great number of advantages:

- More efficient production, with a **lower cost** and a greater quantity of food collected.

- **Sustainability** By having a greater amount of data and better management of them, it is possible to make a more efficient use of pesticides and fertilizers.
- **Access to data.** As all the information is digitized, it is possible to access the necessary data and monitor the status of the crops almost immediately.
- **Greater speed when completing tasks**. Automation and data management allows the various tasks to be carried out more quickly and efficiently.

CHALLENGES AND OPPORTUNITIES

In the agricultural sector, IoT devices are projected to increase from 45 million in 2020 to 75 million by 2025. The application of modern technologies in agriculture has become the basis for the creation of intelligent agricultural infrastructure and has led to an increase in productivity and productivity in the industry. IoT technology is an important factor in the development of knowledge in crop care, assessment of the situation in the cultivation of agricultural products, the right decision-making and automation.

Recent research on IoT in agriculture has focused on addressing the challenges of large-scale farming in the entire supply chain in the field of agro-technologies. Some of these issues are new business models, security, privacy, weather, monitoring and other relevant research in data management smart agriculture has been carried out mainly using wireless sensor technologies.

Evolution of wireless communication

We are actively aware of the promising new generation wireless high-speed data transmission networks around the world, and that IoT technology is also evolving. «G» means «generation», so we will consider the technology «G» fig. 4.

5G technology: In April 2019, technologies for 5G networks were deployed in South Korea (the first country to launch 5G commercial services), Switzerland, and China. The main manufacturer of the equipment is the Chinese company Huawei. "The main task for the fifth generation networks will be to expand the range of frequencies used and increase the capacity of IoT networks.

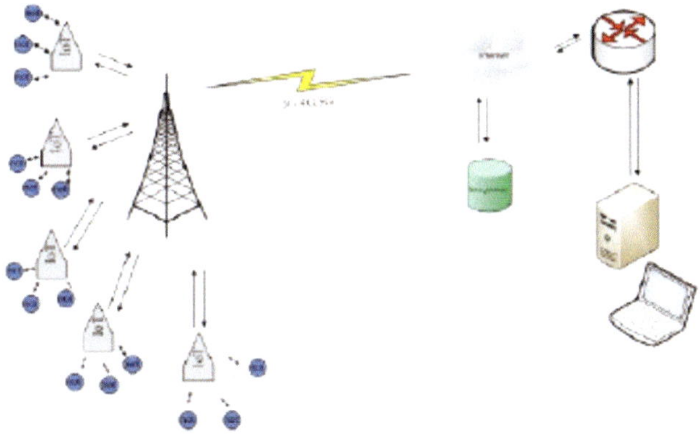

Fig. 4: IoT and G technologies

APPLICATION OF IOT IN AGRICULTURE

There are several examples of the use of IoT in agriculture. Examples include crop and livestock, machinery, irrigation and water quality monitoring, weather monitoring, soil monitoring, disease and pest control, and crop monitoring without such use.

The application of IoT was based on the following functions:

- monitoring
- observation
- Agricultural machinery
- mental agriculture

A. **Monitoring: Several factors can be observed in agriculture**. These factors depend on the agricultural sector. The main factors to be observed are highlighted and discussed as follows.

   - In addition, timely and accurate weather forecasting data, such as climate change and precipitation, can improve yields, such data can help farmers at the planning stage and reduce labor costs. Farmers can take corrective and preventive measures based on the information provided. It can be used to collect pest movement data and remotely control pests on farmers

B. **Monitoring and tracking.** IoT can activate and provide information to agricultural companies to make better decisions.

   - Inspection and tracking of agricultural products allows the consumer to know the complete history of the product, as well as increase confidence in product safety, health issues [9]. GPS tracking allows you to collect product data and link the store to a database to bring the product stock remotely with tracking. Monitoring and tracking allows multiple data to be collected across the supply backup so that the consumer can see the collateral from a full distance.

C. **Greenhouse production**. At the present time, regardless of the time of year, agricultural products are grown because of the production of warm rooms using new technologies, these plants are grown under control.

It allows you to take advantage of any plant by growing it anywhere, anytime, creating appropriate environmental conditions. A lot of work has been done on the use of wireless sensors in the greenhouse to monitor environmental conditions. Recent work with the use of IoT:

- Human resource reduction
- Energy saving

- Improving the efficiency of greenhouse monitoring
- Allows customers to connect directly to the greenhouse farmers.

Layer 1 (the perception layer): At the level of perception, we encounter technologies such as wireless nodes with less processing power and storage capacity. Typically, a wireless sensor assembly consists of a processing module, usually a low-power microcontroller unit (MCU), one or more sensor modules (built-in or external analog or digital sensor devices) and radio frequency communication module, usually supports lowpower wireless communication technologies.

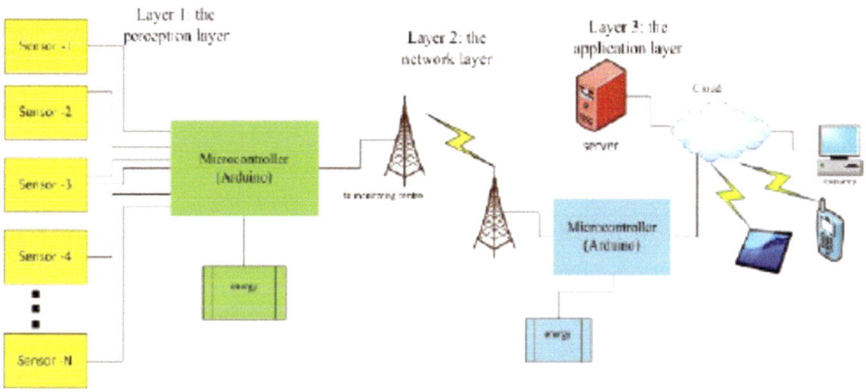

Fig. 5: Obtaining data using IoT devices

Layer 2 (the network layer): At the second level, wireless sensors interact with physical objects and/or their environment, communicate with its neighbors or gateway, creating networks through which data is usually sent to a remote infrastructure for storage, further analysis, processing and dissemination of valuable knowledge that can to be retrieved. Regarding wireless communication communications created a large scientific literature in sensor networks, solving several problems such as energy efficiency, network functions, scalability and reliability.

Layer 3 (the application layer): The application tier is the third tier of the Internet of Things. This is of high importance and, in many ways, this is what the implementation of the Internet of Things contributes. The application layer faces several issues that need to be addressed, such as identifying the device as unique entities. Identifying and manipulating billions of devices around the world will provide direct, Internet access and control over the Internet of the future. Use of software and hardware as it is technology independent when it comes to implementing services. Future interconnected IoT devices in agriculture may include sensors connected to the network.

Appliances and vehicles, weather stations, internet gateways, network storage, RFID scanners, smartphones, tablets, wearable devices and many other devices.

IOT OPPORTUNITIES IN AGRICULTURE

There are a number of benefits to be gained from the use of IoT in agriculture:

- Community farming: The use of IoT can help promote community farming, especially in rural areas. It is possible to promote services that allow to have common data storage, data and information exchange, increase cooperation between farmers and agricultural specialists. Mobile applications and IoT facilities can also be integrated within the community through free or paid services through the use of equipment.
- Safety control and fraud prevention: The challenge in agriculture is not limited to the ability to provide safe and nutritious food, not just pests and pests. Food fraud poses a health challenge and can have negative economic impact Discussion of food fraud components Some product and data integrity can be addressed using IoT technology. IoT can be used to ensure food quality.
- Competitiveness: Increased demand for food products and the use of innovative technology are expected to make the agricultural sector very competitive. It also opens up a sales, monitoring, and marketing direction that allows agriculture to collect data using IoT.
- Cost reduction and benefit: One of the perceived advantages of IoT is the ability to remotely monitor devices and equipment. Applying IoT in agriculture can save time and money when inspecting large areas. The ability to know when and where to use IoT reduces costs and waste.
- Awareness: Information on markets, prices, services in the field of IoT agriculture can be accessed through mobile applications. Also, the implementation of public services and regulatory standards related to various farm production is rapidly becoming available. In addition, consumers interested in new products can easily find farmers or be alerted when new products are available.
- Management: IoT provides real-time monitoring of farm assets and machinery against theft, is used to replace parts and maintain regular crop products in a timely manner.

PROBLEMS OF IOT IN AGRICULTURE

There are several problems associated with the use of IoT. Some problems are lack of security and privacy, data convergence and mutual flexibility, uncertainty in business models.

- **Devices**: There are several costs associated with IoT in agriculture, which can be divided into adjustment costs and operating costs. Setup costs include the purchase of hardware (IoT devices, gateways, base station infrastructure). Operating costs include the use of centralized services or IoT platforms, which allow the management of IoT devices, the exchange of information among other services. Other additional costs stem from the exchange of data between IoT devices, gateway and cloud server, power and service.

- **Entrepreneurship models**: Farmers are interested in entrepreneurship - there are models that support income generation from data collected from their farms using IoT technologies. The information provided is exploited by IoT service providers and this remains a disputed area by farmers for control and ownership of their data.
- **Lack of sufficient knowledge**: Lack of sufficient knowledge is a major factor slowing down the adoption of IoT in agriculture, especially among farmers living in rural areas. It is common for many of these farmers to be unaware of modern technologies evolving where they are found in rural areas and mostly uneducated. The farmer's inability to access information can be a major barrier in the absence of human intervention
- **Interference**: The placement of large IoT devices for agricultural and other purposes, especially using unlicensed spectrum such as ZigBee causes interference with IoT devices. This intervention can lead to data loss and reduce the reliability of the IoT ecosystem [.
- **Reliability**: IoT devices are expected to be placed outdoors. This causes the devices to be exposed to harsh environmental conditions, which can lead to the breakdown of distributed sensors over time and communication failures. The physical safety of distributed IoT sensors and systems must be ensured to protect valuable equipment from severe weather conditions such as floods and storms.
- Nodes and networks: Billions of IoT devices are expected to be deployed in the agricultural sector. Existing gateways and networks will need to support a large number of IoT devices nodes and networks. This requires a smart IoT management system for identification numbers for each node and network.

GLOSSARY

Accuracy (of GPS receivers): The measure of closeness of an object's actual (true) position to the position obtained with a GPS receiver. Accuracy levels are used to rate the quality of GPS receivers.

Active ingredient: In any pesticide product, the component that kills, or otherwise controls, target pests. Pesticides are regulated by the EPA primarily on the basis of active ingredients. See also Inert ingredient.

Aerial Imagery: Photos taken from airplanes, satellites, or UAVs to assist growers to determine variations within an area of interest such as a field.

Aerial remote sensing images: Photographs taken from the air, usually done by drones, to help farmers determine the status of the field.

Agribusiness: Agriculturally related businesses that supply farm inputs (such as fertilizer or equipment) or are involved in the marketing of farm products (such as warehouses, processors, wholesalers, transporters, and retailers). Farms are not usually included when the term agribusiness is used.

Agricultural chemicals: Chemicals used in agriculture, including fertilizers, herbicides and pesticides.

Agricultural data: Data from farm activities and conditions. Data can include soil analysis, nutrition information, seed selection, plant population, and yield data.

Agricultural pollution: Wastes, emissions, and discharges arising from farming activities.

Agricultural product: The term agricultural products is often used as a collective expression and like agricultural commodity, is defined in different ways to meet different circumstances. Agricultural protection zoning: Local zoning codes that include provisions such as large lot size requirements and use limitations to separate farming and related activities from other land uses

Agricultural production: A series of agricultural activities and the resulting agricultural products will be sold in the market.

Agriculture 4.0: An analogy to Industry 4.0. It is the integrated internal and external networking of farming operations as a result of the emergence of smart technology in agriculture.

Agro-chemicals: Chemicals used in agriculture, which include fertilizers, herbicides, and pesticides.

Agronomic data: Data derived from activities and conditions on farm fields. The data can include soil analysis, nutrient information, hybrid selection, plant populations, and yield data.

Agronomy: The science of field-crop production and soil management.

Air pollution: Contamination of the atmosphere by substances that, directly or indirectly, adversely affect human health or welfare

Algorithm: A mathematical formula that can be used to control variable rate applications.

Animal biologics: Vaccines, bacterins, antigens, diagnostic kits and other products of biologic origin.

Aquifer: An underground geological formation, or group of formations, containing usable amounts of groundwater that can supply wells or springs for domestic, industrial, and irrigation uses.

Arid: A relatively dry climate in which annual precipitation is less than 10 inches, which generally is insufficient for crops to be grown without irrigation. Such areas usually are the focus of debate over federal water policies.

Artificial intelligence: Computers simulate intelligent behaviour, such as decision making. AI can refer to anything from computer programs that play chess to speech recognition systems like Amazon Alexa, which are used to interpret and respond to speech.

Auto-Steer: a GPS guidance system that steers agricultural equipment with centimeter accuracy. This level of accuracy requires real time kinematic (RTK) correction of GPS signals. Auto-steer is an add-on component for equipment. It includes both the GPS system to receive and process the signals, software and hardware to allow the input of control maps and the mechanical equipment to actually steer the tractor. Some new tractors are available "auto-steer ready."

Base Map: a simple map that shows the boundaries of a field or section and information about any unique feature (sinkholes, or streams).

Base Station: The RTK-GPS receiver and radio that are placed in a stationary position, functioning as the corrections source for roving tractor units in an area. These stations can be either portable or permanently installed systems and their coverage can range from 5 to 10 miles depending on topographic conditions, antenna height, and radio-transmit power.

Big Data Analytics: Examine a large number of different data sets (called big data) to reveal hidden patterns, unknown correlations, market trends, and customer preferences that can be used for decision making.

Big Data: Information collected in a large number of different data sets through IoT sensors and devices.

Biofuels: Fuels made from biomass, which in the United States largely include corn-based ethanol (blended into gasoline and called gasohol) and soybean-based biodiesel. Biofuels are a subset of renewable fuels, which are a subset of alternative fuels.

Biological control: The practice of using beneficial natural organisms to attack and control harmful plant and animal pests and weeds is called biological control, or biocontrol.

Biological monitoring: Using living organisms to test the quality of either effluent to be discharged into receiving waters, or waters downstream from a discharge.

Biological oxygen demand (BOD): See Biochemical oxygen demand.

Biologics: Immunization vaccines, bacterins, antigens, and antitoxins and other preparations made from living organisms and their products, intended for use in diagnosing, immunizing, or treating humans or animals, or in related research. The Animal and Plant Health Inspection Service has responsibility for approving some animal biologics. See Veterinary biologics.

Biopesticide: A pesticide that is biological in origin (i.e., viruses, bacteria, pheromones, natural plant compounds) in contrast to synthetic chemicals. Transgenic Bt cotton and corn are biopesticides because *Bacillus thuringiensis* (Bt) is a naturally occurring soil bacterium that has been genetically engineered into the plants.

Biopharmaceutical (biopharm) crops: Crops that are bioengineered to produce pharmaceuticals, industrial chemicals and other medical and industrial products (sometimes called plant-made pharmaceuticals).

Biotechnology: Broadly defined as transferring genes from one living entity (plant, animal, or microbe) into another (or into a synthetic compound) using advanced recombinant DNA technology (also called genetic engineering (GE)).

Boom/section Controller: An electronic device that is capable of turning on/off sections of a chemical application boom manually or automatically in combination with GPS positioning and area mapping.

CAN-Bus (in tractors and implements): CAN-Bus is a high-speed, wired data network connection between electronic devices. The hardware/wiring of CAN-Bus networks are generally the same, while the protocols for communication can be different and vary depending on the industry where they are used. These networks are used to link multiple sensors to an electronic controller, which can be linked to relays or other devices on a single set of wires. This reduces the amount of wires needed for a system and allows for a cleaner way to connect additional devices as system demands change.

Contour Line: a line used to represent the same value of an attribute (elevation or yield).

Contour Map: yield map that combines dots of the same intensity/yield level by interpolating (or kriging).

Contouring: interpolation method used to distinguish between different levels of an attribute (elevation, fertility, yield).

Control Segment: the network of tracking stations that monitor and control GPS satellites.

CORS (network): Continuously Operating Reference Station. A network managed by the U.S. office of National Ocean Service (NOAA) to provide GNSS data consisting of carrier phase measurements throughout the United States.

Data Layer (in GIS): A layer of information on a GIS map. A map can have many layers to present different types of information. For example, the first layer of a map may be a satellite image of an area. The next layer may have only lines that represent roads or highways. The next layer may contain topographic information and so forth.

Database: a collection of different pieces of georeferenced information (yield, soil type, fertility) that can be manipulated (layered) in a GIS model.

Differential Correction: correction of a GPS signal that is used to improve its accuracy (to less than 100 m/~330 ft) by using a stationary GPS receiver whose location is known. A second receiver computes the error in signal by comparing the true distance from the satellites to the GPS measured distance.

Differential Global Positioning System (DGPS): a method of using GPS which attains the position accuracy needed for precision farming through differential correction.

Flow Sensor: a sensor that measures the amount of flow through an enclosure (tube, pipe or housing) per unit of time.

Geographic Data: data that contains information about the spatial location (position) and the attribute being monitored (yield, seed population, etc.). Also referred to as spatial data.

Geographic Information System (GIS): a computer-based system used to input, store, retrieve, and analyze geographic data sets. The GIS is usually composed of map-like spatial representations called layers which contain information on a number of attributes such as elevation, land ownership and use, crop yield and soil nutrient levels.

Georeferencing: the process of adding geographic data to yield data or other field attributes either in real-time (on-the-go) or by post-processing or the process of associating data points with specific locations on the earth's surface.

Geo-Stationary Satellite: an orbital path of a satellite that is synchronized with the earth's orbit or space vehicles in an orbit which keeps them over the same location on the earth at all times.

Global Positioning System (GPS): A system using satellite signals (radio-waves) to locate and track the position of a receiver/antenna on the Earth. GPS is a technology that originated in the U.S. It is currently maintained by the U.S. government and available to users worldwide free of charge. There are 30 satellites in the GPS constellation.

Global Positioning System: A system that uses satellite signals to locate and track the position of the receiver/antenna on the Earth.

Global warming: See Climate change and/or Greenhouse gases.

GLONASS (GLObal naya NAvigatsionnaya Sputnikovaya Sistema): Russian version of the American GPS satellite system. It is a radio-based satellite navigation system operated for the Russian government by the Russian Space Forces with a constellation of 24 operational satellites in 2010.

GNSS (Global Navigation Satellite System): Is the standard generic term for satellite navigation systems that provide geo-spatial positioning with global coverage using time signals transmitted from satellites. The United States GPS and the Russian GLONASS are the only two fully operational GNSS. Top of the line GNSS receivers can communicate with both GPS and GLONASS satellites effectively doubling the available reference satellites at any given time.

GPS Antenna: The device that receives satellite signals from space. On most hand-held GPS devices, the antenna is integrated into the receiver device. For machine GPS systems, the antenna is typically an external device that can be mounted on top of the vehicle, away from the receiver.

Grid map: Soil or plant samples can be obtained to analyze predetermined location information in the field. This information can be used to assess fertility requirements and to determine the approximate location of different fertilizer applications.

Grid Mapping: Predetermined locations in a field where soil or plant samples may be obtained for analysis. The test information can be used for assessing fertility needs and determining approximate locations for varying fertilizer and lime applications.

Grid Sampling: the collection of samples from small, uniform-sized cells based on a systematic grid laid out across a field. Grid location in the field is used to develop a field map for the attribute measured.

Ground Control Points: stationary objects/areas on the earth's surface that provide georeferenced points in a remote sensing image/aerial photograph.

Interpolation: mathematical procedure for estimating unknown values from neighboring known data.

ISOBUS: ISOBUS standard 11783 is a communication protocol for the agricultural industry that is used to specify a serial data network for control and communications on forestry or agricultural tractors and implements. ISOBUS-compliant tractors and implements come with round 9-pin connectors.

Kriging: an interpolation method for obtaining stastically unbiased estimates for field attributes (yield, nutrients, elevation) from a set of neighboring points.

LANDSAT (LAND SATellite): US satellites used to study the earth's surface using remote sensing techniques.

Latitude: A global standard coordinate used to identify a position on earth given in degrees, minutes and seconds, indicates the north/south position above/below the equator, positive is in the northern hemisphere and negative is in the southern hemisphere.

Lightbar (in machine guidance): A device connected to a GPS receiver typically consisting of a row of led lights to provide the tractor operator with a visual guide, day or night. The lightbar does not automatically steer the tractor or machine, rather it aids the operator in driving on the imaginary reference line.

Longitude: A global standard coordinate used to identify a position on earth given in degrees, minutes and seconds, indicates the east/west position around the globe from a reference point which overlays Greenwich, England. Negative values are east of Greenwich and positive values are west.

Machine data: Information about how the device works. This can include fuel consumption, machine health indicators, diagnostic codes, and engine performance.

Machine learning: Machine learning is a sub-area of artificial intelligence. The core principle is that the machine acquires data and learns by itself.

NAVSTAR (NAVigation by Satellite Timing And Ranging): the U.S. based global navigation satellite system that was funded by taxpayers and controlled by the DOD.

NDVI Image: The Normalized Difference Vegetation Index (NDVI) is a simple graphical indicator that can be used to analyze remote sensing measurements and assess whether the target being observed contains live green vegetation or not.

Phenome: The set of physical and biochemical traits belonging to an organism.

Phenomics: The measurement of phenomes.

Precision Farming: managing crop production inputs (seed, fertilizer, lime, pesticides, etc.) on a site-specific basis to increase profits, reduce waste and maintain environmental quality.

Prescribed Application: The dispensing of a material or chemical into the field on a prescribed or predetermined basis. A prescription map is generated by an expert (grower and/or agronomist) based on information about the field in use before an application. The prescription determines how much of something will be applied.

Prescription File: A computer generated GIS file that assigns a value to a given geographical area. Example: Nitrogen application rate.

Raster Format: format for storing GIS spatial data in which the data is stored in cells which are addressed by rows and columns of the cells.

Rate Controller: An electronic device that varies the amount of chemical/plant nutrient applied to a given area.

Real-Time Correction: correction of a GPS signal by simultaneously transmitting the differential correction information to a mobile receiver.

Real-Time Kinetic (RTK): procedure whereby carrier-phase corrections are transmitted in real time from a reference receiver to the user's receiver. Depending on local availability, RTK corrections can be delivered by radio modem from an on-site base station or a state's CORS network, or even over the internet using Wi-Fi.

Receiver (in GPS hardware): A computer-radio device that receives satellite information by way of radio waves to determine the position of its antenna relative to the earth's surface. The antenna can be integrated into the receiver or connected externally with a cable.

Remote sensing: Scanning an object by satellite or high-flying aircraft in order to obtain information without direct contact between the sensor and the object.

Satellite: A communications vehicle orbiting the Earth. Satellites typically provide a variety of information from weather data to television programming. Satellites send time-stamped signals to GPS receivers to determine the position on the Earth.

Semi-variance: a measure of the difference between two neighboring data points (defined as 1/2 the square of the difference between two values).

Site-Specific Crop Management (SSCM): The use of yield maps, grid sampling and other precision tools to manage the variability of soil and crop parameters and aid decisions on production inputs.

Soil Map: a map that indicates differences in soil properties (texture, fertility, organic matter, pH, etc.) within a field.

Spatial Data: Data that contains information about the spatial location (position) and the attribute being monitored such as yield, soil properties, plant variables, seed population, etc. Synonymous with geographic data.

Spatial Resolution: the size of the smallest unit that can be identified by a remote sensing observation.

Spatial Variability: differences in observed attribute(s) that are noted between locations in a field.

Spectral Resolution: the ability of sensing system to differentiate between electromagnetic radiation of different wavelengths.

Speed Sensors: sensors that measure the rotational speed of a shaft or the reflection of radio or sound waves off the ground to determine forward speed.

Sustainable intensification: The goal of agriculture is to increase the production of existing farmland while minimizing damage to the environment, conserving biodiversity, and protecting the land for future crops.

UAV: Stands for Unmanned Aerial Vehicle and is an aircraft piloted by remote control or onboard computers. Commonly known as a drone.

Unmanned Aerial Vehicles (UAVs): An unmanned aerial vehicle (UAV), commonly known as a drone and also referred by several other names, is an aircraft without a human pilot aboard. The flight of UAVs may be controlled either autonomously by onboard computers or by the remote control of a pilot on the ground or in another vehicle. In agriculture, UAVs are typically used to survey crops. The available two types of UAVs: fixed-wing and rotary-wing: are both equipped with cameras and are guided by GPS. They can travel along a fixed flight path or be controlled remotely.

Variable irrigation: An emerging technology used in conjunction with a central hub irrigation system.

Variable-Rate Application (VRA): adjustment of the amount of crop input such as seed, fertilizer, lime or pesticides to match conditions (yield potential) in a field.

Variable-Rate Technology (VRT): system of sensors, controllers and agricultural machinery used to perform variable-rate applications of crop production inputs.

Vector Format: a format for storing and displaying GIS spatial data that is stored as points, lines or areas to create a map object. By using a nearly continuous coordinate system, vector data can be more accurately georeferenced than raster data.

Wireless Communication: Data transfer and voice communications using radio frequencies or infrared light.

Yield Map: a map that indicates differences in crop yield within a field. Data is usually collected by a yield monitor on a combine over one to three second intervals.

Yield Monitor: electronic device that continuously measures and records crop yields and moisture on-the-go.

Zone Management: The information-based division of large areas into smaller areas for site specific management applications.

BIBLIOGRAPHY

Allan S, Shi B, Adkins SW. Impact of parthenium weed on human and animal health. Parthenium Weed Biol Ecol Manag. 2018;7:105.

Appleby AP, Müller F, Carpy S. Weed control. In Ullmann's Encyclopedia of Industrial Chemistry, (Ed.). https://doi.org/10.1002/14356007.a28_165

Aqeel-ur-Rehman and Z. A. Shaikh, "Smart agriculture," in *Application of Modern High-Performance Networks*, pp. 120–129, Bentham Science Publishers Ltd, 2009.

Arad, B., Balendonck, J., Barth, R., Ben-Shahar, O., Edan, Y., Hellström, T., et al. (2020). Development of a sweet pepper harvesting robot. *J. Field Robot.* 13–1. doi: 10.1002/rob.21937

Arad, B., Kurtser, P., Barnea, E., Harel, B., Edan, Y., and Ben-Shahar, O. (2019). Controlled Lighting and Illumination-Independent Target Detection for Real-Time Cost-Efficient Applications. The Case Study of Sweet Pepper Robotic Harvesting. *Sensors* 15–1 ,19. doi: 10.3390/s19061390

Arefi, A., Motlagh, A. M., Mollazade, K., and Teimourlou, R. F. (2011). Recognition and localization of ripen tomato based on machine vision. *Aust. J. Crop Sci.* –1144 5 1149.

Bac, C. W., Hemming, J., and van Henten, E. J. (2014a). Stem localization of sweet-pepper plants using the support wire as a visual cue. *Comput. Electr. Agricult.* 120–111 ,105.

Bac, C. W., Henten, E. J., Hemming, J., and Edan, Y. (2014b). Harvesting robots for high-value crops: state-of-the-art review and challenges ahead. *J. Field Robot.* 31:888e911.

Bargoti, S., and Underwood, J. P. (2017). Image Segmentation for Fruit Detection and Yield Estimation in Apple Orchards. *J. Field Robot.* 1060–1039 ,34.

Barriuso, A., G. Villarrubia González, J. de Paz, Á. Lozano, and J. Bajo, "Combination of multi-agent systems and wireless sensor networks for the monitoring of cattle," *Sensors*, vol. 18, no. 2, p. 108, 2018.View at: Publisher Site | Google Scholar

Barth, R., Hemming, J., and Van Henten, E. J. (2019). Angle estimation between plant parts for grasp optimisation in harvest robots. *Biosyst. Eng.* 46–26 ,183.

Basnet, B. and J. Bang, "The state-of-the-art of knowledge-intensive agriculture: a review on applied sensing systems and data analytics," *Journal of Sensors*, vol. 2018, Article ID 3528296, 13 pages, 2018.

Bechar, A. (2010). Robotics in horticultural field production. *Stewart Postharvest Rev.* ,6 11–1. doi: 10.1111/plb.12914

Birrell, S., Hughes, J., Cai, J. Y., and Iida, F. (2019). A field-tested robotic harvesting system for iceberg lettuce. *J. Field Robot.* 21–1 ,37. doi: 10.1002/rob.21888

Blas, M. R., and Blanke, M. (2011). Stereo vision with texture learning for fault-tolerant automatic baling. *Comput. Electr. Agricul.* 168–159 ,75.

Blok, P. M., van Boheemen, K., van Evert, F. K., IJsselmuiden, J., and Kim, G. (2019). Robot navigation in orchards with localization based on Particle filter and Kalman filter. *Comput. Electr. Agricult.* 269–261 ,157.

Borghi, E., Avanzi, J.C., Bortolon, L., Junior, A.L. and Bortolon, E.S. 2016. Adoption and use of precision agriculture in Brazil: Perception of growers and service dealership. *Journal of Agricultural Science* 8 (11): 89.

Boselin, P.S.R., Pradeep, M. and Gajendran, E., 2017. An analysis of smart irrigation system using wireless sensor network. *Star International Journal* 5 3(3): 12-17.

Bramley, R., and Trengove, S.A.M., 2013. Precision agriculture in Australia: Present status and recent developments. *Engenharia Agrícola* 33 (3): 575-588.

Bréant, D., Jézéquel, K. and Lebeau, T., 2002. Optimization of the cell release from immobilized cells of Bacillus simplex cultivated in culture media enriched with Cd2+: influence of Cd2+, inoculum size, culture medium and alginate beads characteristics. *Biotechnology Letters* 24: 1237–1241.

Brown, G. K., and Schertz, C. E. (1967). Evaluating Shake Harvesting of Oranges for the Fresh Fruit Market. *Trans. ASAE* 578–577 ,10.

Buemi, F., Massa, M., Sandini, G., and Costi, G. (1996). The AGROBOT project. *Adv. Space Res.* 189–185 ,18.

Bulanon, D. M., Burks, T. F., and Alchanatis, V. (2010). A multispectral imaging analysis for enhancing citrus fruit detection. *Environ. Control Biol.* 91–81 ,48.

Bulanon, D. M., Kataoka, T., Okamoto, H., and Hata, S. (2004). "Development of a real-time machine vision system for the apple harvesting robot," in *Proceedings of the SICE Annual Conference in Sapporo, Hokkaido Institute of Technolgy*, Japan, 595–598.

Bulanon, D. M., Kataoka, T., Ota, Y., and Hiroma, T. (2002). AE—automation and emerging technologies: a segmentation algorithm for the automatic recognition of Fuji apples at harvest. *Biosyst. Eng.* 412–405 ,83.

Butzen, S., 2013. Common nitrogen fertilizers and stabilizers for corn production. In: Crop Insights. vol. 23.

Ceres, R., Pons, J. L., Jiménez, A. R., Martín, J. M., and Calderón, L. (1998). Design and implementation of an aided fruit-harvesting robot (Agribot). *Indus. Robot* ,25 346–337.

Chen, S. W., Shivakumar, S. S., Dcunha, S., Das, J., Okon, E., Qu, C., et al. (2017). Counting apples and oranges with deep learning: a data-driven approach. *IEEE Robot. Automat. Lett.* 788–781 ,2.

Chen, Y., Lee, W. S., Gan, H., Peres, N., Fraisse, C., Zhang, Y., et al. (2019). Strawberry yield prediction based on a deep neural network using high-resolution aerial orthoimages. *Remote Sens.* 11:1584.

Comba, L., Biglia, A., Aimonino, D. R., and Gay, P. (2018). Unsupervised detection of vineyards by 3D point-cloud UAV photogrammetry for precision agriculture. *Comput. Electr. Agricult.* 95–84 ,155.

Cubero, S., Aleixos, N., Albert, F., Torregrosa, A., Ortiz, C., García-Navarrete, O., et al. (2014). Optimised computer vision system for automatic pre-grading of citrus fruit in the field using a mobile platform. *Precision Agricult.* 94–80 ,15.

d'Grand, E., Rabatel, A. G., Pellenc, R., Journeau, A., and Aldon, M. J. (1987). Magali: A self-propelled robot to pick apples. *Am. Soc. Agricult. Eng. Paper* 358–353 ,46.

Daudelin, J., Jing, G., Tosun, T., Yim, M., Kress-Gazit, H., and Campbell, M. (2018). An integrated system for perception-driven autonomy with modular robots. *Sci. Robot.* 6–1 ,3.

De la Concepcion, A. R., R. Stefanelli, and D. Trinchero, "A wireless sensor network platform optimized for assisted sustainable agriculture," in *IEEE Global Humanitarian Technology Conference (GHTC 2014)*, pp. 159–165, San Jose, CA, USA, October 2014.View at: Publisher Site | Google Scholar

Dey, D., Mummert, L., and Sukthankar, R. (2012). "Classification of plant structures from uncalibrated image sequences," in *Proceedings of the 2012 IEEE Workshop on the Applications of Computer Vision (WACV)*, Breckenridge, CO, 329–336.

Edan, Y., Han, S., and Kondo, N. (2009). *Automation in Agriculture, Springer Handbook of Automation.* Berlin: Springer Berlin Heidelberg, 1128–1095.

Edan, Y., Rogozin, D., Flash, T., and Miles, G. E. (2000). Robotic melon harvesting. *IEEE Trans. Robot. Automat.* 835–831 ,16.

Ehud, B., Rotem, M., and Ohad, B. (2016). Colour-agnostic shape-based 3D fruit detection for crop harvesting robots. *Biosyst. Eng.* 70–57 ,146.

Fischer, J. W., Greiner, K., Lutman, M. W., Webber, B. L., and Vercauteren, K. C. (2019). Use of unmanned aircraft systems (UAS) and multispectral imagery for quantifying agricultural areas damaged by wild pigs. *Crop Protect.* 125:104865.

Fu, L., Feng, Y., Majeed, Y., Zhang, X., Zhang, J., Karkee, M., et al. (2018). Kiwifruit detection in field images using Faster R-CNN with ZFNet. *IFAC Papers On Line* ,51 50–45.

Gago, J., C. Douthe, R. E. Coopman et al., "UAVs challenge to assess water stress for sustainable agriculture," *Agricultural Water Management*, vol. 153, pp. 9–19, 2015.

Ivanova, E., Teunou, E. and Poncelet, D., 2005. In: Gruev, B., Nikolova, M., Donev, A. (Eds.), Alginate based macrocapsules as inoculants carriers for production of nitrogen biofertilizers. Proceedings of the Balkan Scientific Conference of Biology in Plovdiv (Bulgaria), pp. 90–108.

Jorquera, M., Herna´ndez, M., Rengel, Z., Marschner, P. and Mora, M.L., 2008. Isolation of culturable phosphobacteria with both phytatemineralization and phosphatesolubilization activity from the rhizosphere of plants grown in a volcanic soil. *Biology and Fertility of Soils* 44: 1025–1034.

Kalubhai, P., 2015. Present Status and Prospects of Micro Irrigation System in Mahisagar District of Gujarat. Doctoral dissertation, Anand Agricultural University, Anand, India.

Kennedy, I.R., Choudhury, A.T.M.A. and Kecske´s, M.L., 2004. Non-symbiotic bacterial diazotrophs in crop-farming systems: can their potential for plant growth promotion be better exploited? Soil Biology and Biochemistry 36: 1229–1244.

Keskin, M., and Sekerli, Y.E., 2016. Awareness and adoption of precision agriculture in the Cukurova region of Turkey. *Agronomy Research* 14 (4): 1307-1320.

Kim, S.Y., Kim, M., Lee, S., Lee, J. and Ko, G., 2012. Survival of microorganisms on antimicrobial filters and the removal efficiency of bioaerosols in an environmental chamber. *Journal of Microbiology and Biotechnology* 22 (9): 1288–1295.

Kumar R, Katiyar R, Kumar S, Kumar T, Singh V. *Lantana camara*: an alien weed, its impact on animal health and strategies to control. *J Exp Biol Agric Sci.* 2016;4:321–37.

Kurtulmus F, Lee WS, Vardar A (2014) Immature peach detection in colour images acquired in natural illumination conditions using statistical classifiers and neural network. *Precis Agric* 15(1):57–79.

Linker R, Cohen O, Naor A (2012) Determination of the number of green apples in rgb images recorded in orchards. *Comput Electron Agric* 81:45–57.

Liu W, Anguelov D, Erhan D, Szegedy C, Reed S, Fu C-Y, Berg AC (2016) Ssd: single shot multibox detector. In: E*uropean conference on computer vision*. Springer, Berlin, pp 21–37.

Maher, A., E. Kamel, F. Enrico, I. Atif, and M. Abdelkader, "An intelligent system for the climate control and energy savings in agricultural greenhouses," *Energy Efficiency*, vol. 9, no. 6, pp. 1241–1255, 2016.V

Mohammed, G., F. Trolard, M. Gillon, A. L. Cognard-Plancq, A. Chanzy, and G. Bourrie, "Combination of a crop model and a geochemical model as a new approach to evaluate the sustainability of an intensive agriculture system," *Science of The Total Environment*, vol. 595, pp. 119–131, 2017.

Naz, M.Y., Sulaiman, S.A., 2016. Slow release coating remedy for nitrogen loss from conventional urea: a review. Journal of Controlled Release 225(10): 109–120.

Neethirajan, S., 2017. Recent advances in wearable sensors for animal health management. Sensing and Bio-Sensing Research 12: 15–29.

Ni, B., Liu, M. and Lu, S., 2009. Multifunctional slow-release urea fertilizer from ethylcellulose and superabsorbent coated formulations. Chemical Engineering Journal 155: 892–898.

Park, D. H. and J. W. Park, "Wireless sensor network-based greenhouse environment monitoring and automatic control system for dew condensation prevention," *Sensors*, vol. 11, no. 4, pp. 3640–3651, 2011.

Park, M., Kim, C., Yang, J., Lee, H., Shin, W., Kim, S. and Sa, T., 2005. Isolation and characterization of diazotrophic growth promoting bacteria from rhizosphere of agricultural crops of Korea. *Microbiological Research* 160: 127–133.

Paustian, M., and Theuvsen, L., 2017. Adoption of precision agriculture technologies by German crop farmers. *Precision Agriculture* 18(5): 701–716.

Perez-Ruiz, M., P. Gonzalez-de-Santos, A. Ribeiro et al., "Highlights and preliminary results for autonomous crop protection," *Computers and Electronics in Agriculture*, vol. 110, pp. 150–161, 2015.

Puri, M., Solanki, A., Padawer, T., Tipparaju, S.M., Moreno, W.A. and Pathak, Y., 2015. Introduction to Artificial Neural Network (ANN) as a Predictive Tool for Drug Design, Discovery, Delivery, and Disposition: Basic Concepts and Modeling; Elsevier Inc.: Amsterdam, the Netherlands, pp. 3–13.

Rakun J, Stajnko D, Zazula D (2011) Detecting fruits in natural scenes by using spatial-frequency based texture analysis and multiview geometry. *Comput Electron Agric* 76(1):80–88.

Stajnko D, Lakota M, Hoèevar M (2004) Estimation of number and diameter of apple fruits in an orchard during the growing season by thermal imaging. *Comput Electron Agric* 42(1):31–42

Universal Robot Support. https://www.universal-robots.com/download/. Accessed 23 Oct 2019.

Xiaofeng, L., Linlin, Q., Linjian, L. and Gang, W. 2015. Design and implementation of modern greenhouse remote monitoring system based on the Android system. In: *34th Chinese Control Conference* (CCC), pp. 5742–5746.

Xu, Z., Chen, J., Wang, Y. and Fan, Z., 2016. A remote monitoring system for greenhouse based on the internet of things. In: MATEC Web of Conferences, EDP *Sciences* 77:1-4

Yadav T, Chopra NK, Chopra NK, Kumar R, Soni PG. Assessment of critical period of crop-weed competition in forage cowpea (*Vigna unguiculata*) and its effect on seed yield and quality. *Indian J Agron.* 2018;63:124–7.

Yan-E, D., 2011. Design of intelligent agriculture management information system based on IoT. In: *Proceedings of the 4th International Conference on Intelligent Computation Technology Automation* ICICTA 2011, Vol. 1, pp 1045–1049

Zahid, M., Abbasi, M.K., Hameed, S. and Rahim, N., 2015. Isolation and identification of indigenous plant growth promoting rhizobacteria from Himalayan region of Kashmir and their effect on improving growth and nutrient contents of maize (*Zea mays* L.). *Frontiers in Microbiology* 6(207): 1–10.

Zhang, X., Zhang, C., Fang, J., Yu, X. and Liang, J., 2009. Smart sensor nodes for wireless soil temperature monitoring systems in precision agriculture. *Transactions of the Chinese Society for Agricultural Machinery* 40: 237-240.

INDEX